# Boundary Elements and other Mesh Reduction Methods XLVII

WIT*PRESS*

WIT Press publishes leading books in Science and Technology.
Visit our website for the current list of titles.
www.witpress.com

WIT*eLibrary*

Home of the Transactions of the Wessex Institute.
Papers published in this volume are archived in the WIT elibrary in volume 136 of WIT Transactions on Engineering Sciences (ISSN 1743-3533).
The WIT electronic-library provides the international scientific community with immediate and permanent access to individual papers presented at WIT conferences.
http://library.witpress.com

# Forty Seventh International Conference on Boundary Elements and other Mesh Reduction Methods

# BEM/MRM 47

**Conference Chairmen**

**Alexander H. D. Cheng**
*University of Mississippi, USA*
*Member of WIT Board of Directors*

**Stavros Syngellakis**
*Wessex Intitute, UK*
*Member of WIT Board of Directors*

**International Scientific Advisory Committee**

Andre Buchau
Weiqiu Chen
Jeng-Tzong Chen
Mario Cvetković
João Batista de Paiva
Petia Dineva
Eduardo Divo
Hrvoje Dodig
Chunying Dong
Ney Dumont
Zhuojia Fu
Alexander Galybin
Xiao-Wei Gao
Luis Godinho
Andreas Karageorghis
Alain Kassab
John Katsikadelis
Edson Leonel
Daniel Lesnic
George Manolis
Ilia Marchevsky
Liviu Marin
Juraj Mužik
Leandro Palermo Jr
Ernian Pan
Dragan Poljak
Jure Ravnik
Antonio Romero Ordóñez
Božidar Šarler
Vladimir Sládek
Elena Strelnikova
Antonio Tadeu
Wolf Yeigh
Jianming Zhang

**Organised by**

*Wessex Institute, UK*

*University of Mississippi, USA*

**Sponsored by**

*WIT Transactions on Engineering Sciences*

# WIT Transactions

Wessex Institute
Ashurst Lodge, Ashurst
Southampton SO40 7AA, UK

We would like to express our thanks to the conference Chairs and members of the International Scientific Advisory Committees for their efforts during the 2024 conference season.

## Conference Chairs

**Alexander Cheng**
*University of Mississippi, USA*
(Member of WIT Board of Directors)

**Stavros Syngellakis**
*Wessex Institute, UK*
(Member of WIT Board of Directors)

## International Scientific Advisory Committee Members 2024


**Socrates Basbas** Aristotle University of Thessaloniki, Greece
**João Batista de Paiva** University of São Paulo, Brasil
**Alma Bojórquez-Vargas** Universidad Autonoma De San Luis Potosi, Mexico
**Colin Booth** University of the West of England, UK
**Carlos Borrego** University of Aveiro, Portugal
**Roman Brandtweiner** Vienna Universiy of Economics and Business, Austria
**André Buchau** University of Stuttgart, Germany
**Paúl Carrión Mero** Higher Polytechnic School of the Litoral, Ecuador
**João Manuel Carvalho** University of Lisbon, Portugal
**Weiqiu Chen** Zhejiang University, China
**Jeng-Tzong Chen** National Taiwan Ocean University, Taiwan
**Mario Cvetković** University of Split, Croatia
**Maria da Conceição Cunha** University of Coimbra, Portugal
**Luca D'Acierno** University of Naples Federico II, Italy
**Pablo Díaz Rodríguez** University of La Laguna, Spain
**Eleni Didaskalou** University of Piraeus, Greece
**Petia Dineva** Bulgarian Academy of Sciences, Bulgaria
**Eduardo Divo** Embry-Riddle Aeronautical University, USA
**Hrvoje Dodig** University of Split, Croatia
**Chunying Dong** Beijing Insitute of Technology, China
**Mauro Dujmović** Juraj Dobrila University of Pula, Croatia
**Ney Dumont** Pontifical Catholic University of Rio de Janeiro, Brazil
**Siyabulela Fobosi** University of Fort Hare, South Africa
**Zhuo Jia Fu** Hohai University, China
**Alexander Galybin** Schmidt Institute of Physics of the Earth, Russia
**Xiao-Wei Gao** Dalian University of Technology, China
**Eric Gielen** Universitat Politècnica de València, Spain
**Emanuele Giorgi** Tecnologico de Monterrey, Mexico
**Luis Godinho** University of Coimbra, Portugal
**Andreas Karageorghis** University of Cyprus, Cyprus
**Alain Kassab** University of Central Florida, USA
**Hiroshi Kato** Hokkaido University, Japan
**John Katsikadelis** National Technical University of Athens, Greece
**Dima Legeyda**, Newcastle University, UK
**Edson Leonel** University of São Paulo, Brazil
**Daniel Lesnic** University of Leeds, UK
**Danila Longo** University of Bologna, Italy
**Isabel Madaleno** University of Lisbon, Portugal
**Robert Mahler** University of Idaho, USA
**Irina Malkina-Pykh** St Petersburg State Institute of Psychology and Social Work, Russia
**George Manolis** Aristotle University of Thessaloniki, Greece
**Ilia Marchevsky** Bauman Moscow State Technical University, Russia
**Liviu Marin** University of Bucharest, Romania
**José Luis Miralles i Garcia** Polytechnic University of Valencia, Spain
**Juraj Mužík** University of Žilina, Slovakia
**Yasuo Ohe** Tokyo University of Agriculture, Japan
**Özlem Özçevik** Istanbul Technical University, Turkey
**Leandro Palermo Jr** State University of Campinas, Brazil
**Ernian Pan** National Yang Ming Chiao Tung University, Taiwan
**Marilena Papageorgiou** Aristotle University of Thessaloniki, Greece
**Marko Perić** University of Rijeka, Croatia
**Filomena Pietrapertosa** National Research Council of Italy, Italy

**Lorenz Poggendorf** Toyo University, Japan
**Dragan Poljak** University of Split, Croatia
**Dimitris Prokopiou** University of Piraeus, Greece
**Elena Cristina Rada** Insubria University of Varese, Italy
**Marco Ragazzi** University of Trento, Italy
**Jure Ravnik** University of Maribor, Slovenia
**Antonio Romero Ordóñez** University of Sevilla, Spain
**Francesco Russo** Mediterranea University of Reggio Calabria, Italy
**Monica Salvia** National Research Council of Italy, Italy
**Božidar Šarler** University of Ljubljana, Slovenia
**Marco Schiavon,** University of Padova, Italy
**Marichela Sepe** DICEA-Sapienza Università di Roma, Italy
**Vladimír Sládek** Slovak Academy of Sciences, Slovakia
**Elena Strelnikova** National Academy of Sciences of Ukraine, Ukraine
**Antonio Tadeu** University of Coimbra, Portugal
**Carlo Trozzi** Teche Consulting srl, Italy
**Sirma Turgut** Yildiz Technical University, Turkey
**Wolf Yeigh** University of Washington, USA
**Jianming Zhang** Hunan University, China
**Meng-Cong Zheng** National Taipei University of Technology, Taiwan

# Boundary Elements and other Mesh Reduction Methods XLVII

**Editors**

**Alexander H. D. Cheng**
*University of Mississippi, USA*
*Member of WIT Board of Directors*

**Stavros Syngellakis**
*Wessex Institute, UK*
*Member of WIT Board of Directors*

WIT*PRESS* Southampton, Boston

**Editors:**

**Alexander H. D. Cheng**
*University of Mississippi, USA*
*Member of WIT Board of Directors*

**Stavros Syngellakis**
*Wessex Institute, UK*
*Member of WIT Board of Directors*

Published by

**WIT Press**
Ashurst Lodge, Ashurst, Southampton, SO40 7AA, UK
Tel: 44 (0) 238 029 3223; Fax: 44 (0) 238 029 2853
E-Mail: witpress@witpress.com
http://www.witpress.com

For USA, Canada and Mexico

**Computational Mechanics International Inc**
25 Bridge Street, Billerica, MA 01821, USA
Tel: 978 667 5841; Fax: 978 667 7582
E-Mail: infousa@witpress.com
http://www.witpress.com

British Library Cataloguing-in-Publication Data

A Catalogue record for this book is available
from the British Library

ISBN: 978-1-78466-493-0
eISBN: 978-1-78466-494-7
ISSN: (print) 1746-4471
ISSN: (on-line) 1743-3533

*The texts of the papers in this volume were set individually by the authors or under their supervision. Only minor corrections to the text may have been carried out by the publisher.*

# Preface

This issue contains papers selected from the 47th International Conference on Boundary Elements and Other Mesh Reduction Methods (BEM/MRM 47). The conference series was founded by Professor Carlos A. Brebbia in 1978, with its first meeting held in Southampton, UK. For the next 46 years, scientists and engineers have used this gathering to exchange the progresses made in the field. The continued success of the meeting is a result of the strength of the research on boundary elements and mesh reduction techniques being carried out all over the world.

This volume collects some of the papers presented at the conference. Advanced methods are presented for the evaluation of parameters in boundary element and boundary integral formulations of solids and fluids. Advances in the free element method and the meshless collocation method are also presented.

The boundary element method (BEM) is combined with a homogenization technique to modelling concrete and with the multiscale technique for modelling concrete plates. It is also applied to the solution of corrosion and acoustic wave scattering problems, in the latter case based on the Burton-Miller method. The impact of stress concentration due to a spherical cavity is accounted for in a BEM formulation yielding near-corner stress values.

Flow in two dimensions is simulated using a boundary integral equation combined with the vortex method; granular debris flow is modelled using the local moving least squares method. In other applications, the solid-solid interface problem is solved using non-dissipative dynamics and the problem of parallel robot car parking using artificial neural networks.

The Editors would like to thank all authors for the quality of their papers and other colleagues for their help in reviewing the material.

The Editors
2024

# Contents

# SECTION 1
# ADVANCED FORMULATIONS AND INVERSE PROBLEMS

# NODE LOCATION OPTIMIZATION AND INQUIRY INTO GEOMETRY-PRESERVING VERSUS ISOPARAMETRIC FORMULATION IN THE COLLOCATION BOUNDARY ELEMENT METHOD

NEY AUGUSTO DUMONT
Department of Civil and Environmental Engineering, Pontifical Catholic University of Rio de Janeiro, Brazil

ABSTRACT

This paper might have a subtitle, '– homothetic node and element generation along boundary patches', with the justification 'to simplify and speed up the numerical simulation'. We have recently laid down the theoretical basis for the consistent formulation of the collocation boundary element method, as it should have been conceived from the beginning. We proposed a convergence theorem for two- and three-dimensional problems of elasticity and potential, which applies to arbitrarily curved elements in the frame of an isoparametric analysis. We also showed that arbitrarily high precision and accuracy may be achieved in the code implementation for two-dimensional problems – limited only by the machine's capacity to represent numbers. On the other hand, there still is the cost–benefit question – considering that the mathematical governing equations are adequate for representing the physical phenomenon – of how to improve a real problem's simulation without increasing the number of degrees of freedom. The first possibility is to increase the polynomial order of the interpolating functions (p-refinement). The second possibility is with local, adaptive, h-refinement of the discretization mesh. We may also attempt to optimize the geometric location – inside a boundary element – to which the problem's primary parameters are attached. Independently of that, an isoparametric formulation may fail to reproduce the exact geometry of the idealized physical problem. Since, for two-dimensional problems, we have the boundary element formulation under control regarding all numerical evaluations, we assess how an isoparametric analysis – with the introduced elegancy of a convergence theorem – compares to a formulation that preserves the problem's idealized geometry but is not isoparametric, in general. We present the conceptual formulation, code implementation, and numerical illustrations that go from the simple case of an infinite plate with a circular hole to very challenging – physically unrealistic but mathematically conceivable – topological applications.

*Keywords: consistent boundary elements, collocation, isoparametric analysis, isogeometric approach, node location optimization, homothetic mesh refinement.*

## 1 INTRODUCTION

Our recent proposition of the 'consistent' collocation boundary element method (CBEM) has required a long maturation time since the primitive proposal [1] of 30 years ago (Here, the word 'primitive' may refer to 'original, primary, crude or rudimentary', but also to 'seminal', as the concept of 'complex singularity poles' – bad marketing but trustful mathematics – was laid down for the first time). Some contributions followed sparsely [2]–[5] until the three-paper publication [6]–[8], which brought the theoretical basis for general linear problems of elastostatics and steady-state potential, with code-implementation and applications for the two-dimensional (2D) case. Although properly using the concept of 'complex singularity poles', these papers were based on real-variable Cartesian coordinates. A fully complex-variable reformulation followed [9], also as the subject of a technical presentation during last year's BEM/MRM conference [10].

Despite the sound mathematics laid down in the papers just outlined, the question 'can we do better?' (for the particular case of 2D linear problems of elastostatics and steady-state potential) persists concerning an efficient numerical simulation:

WIT Transactions on Engineering Sciences, Vol 136, © 2024 WIT Press
www.witpress.com, ISSN 1743-3533 (on-line)
doi:10.2495/BE470011

1. A first issue refers to an eventual optimal node location inside a boundary element.
2. An efficient – problem-dependent – mesh-refinement technique should be considered.
3. Moreover, the isoparametric formulation – and the related, elegant convergence theorem proposed in Dumont [6] for generally curved elements – may require the approximate representation of the problem's boundary geometry, and, then, not be the best one.

We are concerned rather with concepts than with code-implementation aspects and numerical assessments. This contribution is meant as the layout of a paper in preparation [11].

The denomination 'geometry-preserving' is in the present context more informative than and not committed to the term 'isogeometric' proposed by Cottrell et al. [12] for widespread applications also in the frame of the boundary element method – which do not take into account the mathematical consistency of our developments.

Before proceeding with the items above, it is necessary to outline a consistent notation [6]–[9], [13], which was restricted to the isoparametric formulation. The following developments are less formal but perhaps more intuitive, as we first resort to two elasticity applications, taken from Dumont [10], [13] and to be seen not just as a repetition but rather as a motivation and the illustration of how the numerical solution possibilities are deemed to evolve.

## 2 TWO ILLUSTRATIVE APPLICATIONS

**Infinite plate with a hole.** Fig. 1 shows on the left an infinite plate with a circular hole of radius $a$. The displacement and stress solutions for a uniform stress field $\sigma_{xx} = 1$, $\sigma_{yy} = \tau_{xy} = 0$ applied at infinity, according to Sadd [14], for instance, are compactly expressed in polar coordinates as

$$\begin{aligned} u_r &= \frac{1}{4Gr^3}\left[\cos 2\theta\left(4(1-\nu)a^2r^2 + r^4 - a^4\right) + (1-2\nu)r^4 + a^2r^2\right] \\ u_\theta &= \frac{1}{4Gr^3}\sin 2\theta\left[4\nu a^2r^2 - \left(a^2+r^2\right)^2\right] \end{aligned} \tag{1}$$

$$\begin{aligned} \sigma_{rr} &= \frac{1}{2}\left(1-\frac{a^2}{r^2}\right) + \frac{1}{2}\left(1-4\frac{a^2}{r^2}+3\frac{a^4}{r^4}\right)(2\cos^2\theta - 1) \\ \sigma_{\theta\theta} &= \frac{1}{2}\left(1+\frac{a^2}{r^2}\right) - \frac{1}{2}\left(1+3\frac{a^4}{r^4}\right)(2\cos^2\theta - 1) \\ \tau_{r\theta} &= \left(-1-2\frac{a^2}{r^2}+3\frac{a^4}{r^4}\right)\sin\theta\cos\theta \end{aligned} \tag{2}$$

The maximum stress value is $\sigma_{\max} = \sigma_{\theta\theta}(r = a, \theta = \pm\pi/2) = 3$, and we also obtain the highest normal stress in the vertical direction $\sigma_{\theta\theta}(r = a, \theta = 0) = -1$.

As indicated on the right in the figure, we discretize the circular cavity with 10 quartic elements and 40 equally spaced, equidistant nodes from the center – not taking advantage of the double symmetry. The main source of errors in this simulation is related to the fact that the circular surface is modeled with quartic polynomial elements, with non-smooth transition between elements. In fact, we measure for nodes 1, 5, 9, 13, 17, 21, 25, 29, 33 and 37 the same lack of smoothness, with $(\theta^+ - \theta^-)/2\pi \approx 0.4999900076$, when it should be exactly $1/2$. Then, we should not expect in our numerical evaluations a relative accuracy error smaller than about 0.00001998. When evaluating results, points along a horizontal axis on the right in Fig. 1 – which goes through nodes 21 and 1 – should present the largest errors, as there is some angularity there. Tangents about nodes 11 and 31 are horizontal and we should expect

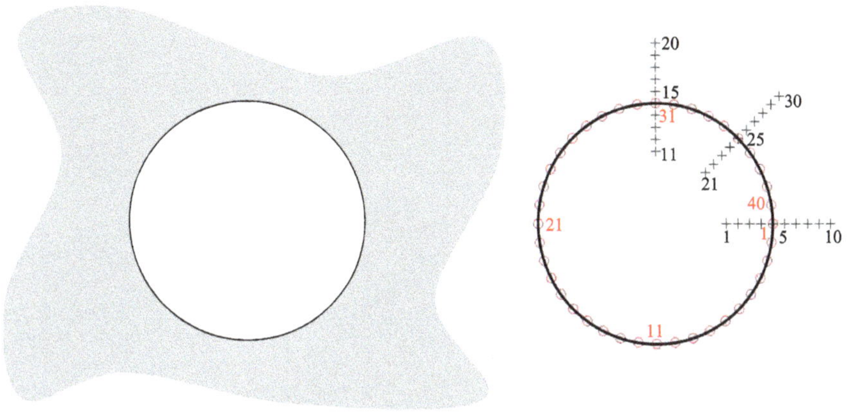


Figure 1: Infinite plate with a circular hole (left) modeled as 10 quartic elements (right) with 30 points at which results are to be evaluated.

the smallest errors measured along a vertical from there. Nodes 6, 16, 26 and 36, which are located at angles multiple of $\pi/4$ of inclination, do not have tangents inclined of exact multiples of $\pi/4$, since they are not middle nodes of the quartic elements. Such lacks of polar symmetry are intentional in order to have geometric simulation errors introduced in our model. The figure also indicates three series of 10 points, each, at which stress results are to be evaluated. Points 1, 11 and 21 are $0.6 + 10^{-10}$ distant from the center, with interpolated points up to the respective points 10, 20 and 30, which are at a distance $1.5 + 10^{-10}$ from the center, in such a way that points 1 ... 4, 11 ... 14 and 21 ... 24 are actually internal to the cavity, that is, external in relation to the open domain of interest. We intentionally set points 5, 15 and 25 inside the open domain and just $10^{-10}$ distant from the respective nodes 1, 31 and 36. Besides the geometry errors, we should expect some round-off errors related to these close points. This is assessed in Dumont [10], [13] and not repeated here.

The superiority of applying the present geometry-preserving formulation to this kind of problem is numerically demonstrated in Dumont [11].

**Topologically extremely challenging problem.** Fig. 2 represents a 2D domain (about 25 units across) with some challenging topological features, to be subjected to a series of elastic fields, as described in Dumont [9] and summarized in Dumont [13], which is on the other hand a development of a simpler numerical model proposed in Dumont [8]. Readers are referred to these papers for the complete description of this *cut-out* test and thorough result assessments. The cusp at node 1 has an internal angle of about $10^{-8}$ *rad*, the external angle at node 17 is of about $10^{-13}$ *rad*, and the strip of material between the cavity and the external boundary is only about $10^{-4}$ unities wide, for an isoparametric formulation with quadratic elements. As detailed in Dumont [9], these angles and distances are much larger when we just switch to quartic elements, which makes the geometric problem discretization-dependent: a disadvantage of the isoparametric formulation, to be addressed next.

The indicated crosses in Fig. 2 are a total of 41 – in part internal and in part external – points at which stress results are to be numerically evaluated for the applied stress fields. Some of these points are very close to the boundary, as described and assessed in Dumont [8], [9]. Most important, we generate between internal point 30 and node 69, which are visually indistinguishable from each other in the figure, a series of 10 points that approach node 69 at geometrically decreasing distances, as indicated in the first row of Table 1 of Dumont [9],

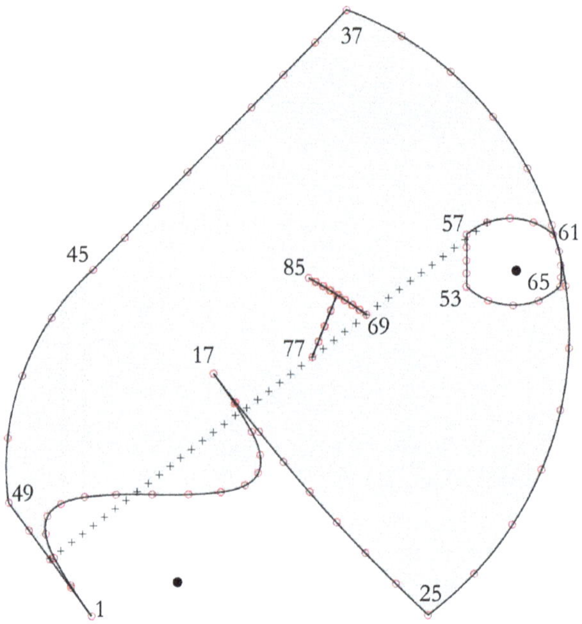


Figure 2: Two-dimensional figure with some challenging topological issues and to be subjected to a series of elastic fields, as reproduced from Dumont [9].

[13]. A similar series of 10 very close points to node 17 is also generated, with the distances indicated in the second row of the mentioned Table: these distances go as small as $10^{-17}$ and $10^{-18}$. If we consider all geometric data given in meters, the smallest distances are actually about one thousandth of a typical proton size, which is meaningless in terms of continuum mechanics but not mathematically!

As indicated in Dumont [13], this elastic body is subjected to two rigid-body translations and a set of four linear, quadratic, cubic, quartic and quintic polynomial fields, thus a total of 22 fundamental (that is, homogeneous) solutions of the elastostatics problem for homogeneous, isotropic material, as given in eqns (15–19) of Dumont [8] in terms of real variables, and very compactly on the right in eqn (34) of Dumont [9], in terms of the complex $z = x + iy$. Rigid-body rotation is obtained as the combination of two of the linear solutions.

The main issue when dealing with such a problem in the frame of an isoparametric formulation is that the challenging geometry is discretization-dependent. Then, distinct meshes reproduce distinct, albeit approximate, problems – and then not strictly comparable.

## 3 BASIC PROBLEM FORMULATION IN THE CONSISTENT, COLLOCATION BOUNDARY ELEMENT METHOD

The peculiarities of present interest for the evaluation of results at internal points are dealt with in the papers referenced above, and shall not be addressed here.

Whether using real or complex variables, the basic system matrix to be solved in the frame of the CBEM has the format

$$\mathbf{H}(\mathbf{d} - \mathbf{d}^p) = \mathbf{G}(\mathbf{t} - \mathbf{t}^p)_{ad} \tag{3}$$

In this equation, **H** is the square, double-layer potential matrix of order $n_d = 2n_n$, for 2D elasticity and the problem discretized with $n_n$ nodal points, and **G** is the single-layer potential matrix with $2n_n$ rows and $2(n_n + n_e)$ columns, as we code for $n_e$ elements of any order $o_e$, in principle taking into account that the left and right tangents at a nodal point connecting two elements are different, in the isoparametric formulation for generally curved boundaries. The number of columns of **G** may be significantly smaller in the patch-related, geometry-preserving formulation we are about to present.

As laid down in Dumont [4], [6], we are assuming just for the sake of elegant and compact formulation that some *particular* solution of interest is known – whether or not related to non-zero body forces – and may be approximately expressed as boundary nodal displacement $\mathbf{d}^p$ and traction $\mathbf{t}^p$ data. This is the case of the infinite plate with a hole of Fig. 1 [10], [13]. The problem's primary boundary displacement and traction parameters are $\mathbf{d}$ and $\mathbf{t}$, which are in part known and in part to be obtained in the frame of a general mixed-boundary formulation ($\mathbf{t} = \mathbf{0}$ for the problem of Fig. 1, and $\mathbf{d}$ is to be obtained). As comprehensively assessed in Dumont [3], [6]–[8], we write for consistency that the traction $(\mathbf{t} - \mathbf{t}^p)_{ad}$ is *admissible*, in equilibrium with the applied domain forces: this follows the same mathematical/mechanical principle that, since – for a finite domain – rigid-body displacement amounts of $(\mathbf{d} - \mathbf{d}^p)$ cannot be transformed into forces, also non-equilibrated forces should not be transformed into displacements (see Section 5.3).

In complex-variable, 2D plane-strain elasticity, the matrices of eqn (3) are [9]

$$G_{s\ell}=\int_\Gamma \begin{bmatrix} -(3-4\nu)\ \ln(z\bar{z}) & \dfrac{z}{\bar{z}} \\ \dfrac{\bar{z}}{z} & -(3-4\nu)\ \ln(z\bar{z}) \end{bmatrix} \frac{|J|_{(at\ \ell)} N_\ell^{o_e} \mathrm{d}\xi}{16\pi G\,(1-\nu)} \tag{4}$$

$$H_{sn}=\int_\Gamma \begin{bmatrix} (3-4\nu)\dfrac{z'}{z}-\dfrac{\bar{z}'}{\bar{z}} & \dfrac{z'}{\bar{z}}-\dfrac{z}{\bar{z}^2}\bar{z}' \\ -\dfrac{\bar{z}'}{z}+\dfrac{\bar{z}}{z^2}z' & \dfrac{z'}{z}-(3-4\nu)\dfrac{\bar{z}'}{\bar{z}} \end{bmatrix} \frac{iN_n^{o_e}\mathrm{d}\xi}{8\pi\,(1-\nu)}+\delta_{sn} \tag{5}$$

Real-variable expressions [7] might substitute for the above, but the complex representation is undeniably simpler. Here, $z = x + iy$, and we use $n = -iz'/|J| \Leftrightarrow \bar{n} = i\bar{z}'/|J|$ for the unit normal, also considering $\mathrm{d}\Gamma = |J|\mathrm{d}\xi$. The material's shear modulus and Poisson's ratio are $G$ and $\nu$.

The rows above refer to the *source*, complex point force $p_s^* = (p_x^* + i\ p_y^*)|_{(at\ s)}$ and its conjugate $\bar{p}_s^* = (p_x^* - i\ p_y^*)|_{(at\ s)}$. The first columns stand for either *node* $n$ or *locus* $\ell$ on a boundary segment, to which either complex displacements $d_n = (d_x + i\ d_y)|_{(at\ n)}$ or tractions $t_\ell = (t_x + i\ t_y)|_{(at\ \ell)}$ are attached.

Only the first rows of the above matrices need to be implemented in a code [9].

We develop eqns (4) and (5) for evaluations using Gauss–Legendre quadrature and then eventually accrue mathematically exact corrections conditioned by three logical constants $<no\_sing,\ sing,\ quasi\_sing>$. This is thoroughly addressed in Dumont [9], [13].

## 4 CONSISTENT NOTATION AND IMPLEMENTATION POSSIBILITIES

We present the basic notation formally proposed in the previous papers [6], [13], but add the possibility of working with the concept of boundary patches and geometry-preserving description. We resort to the applications of Section 2 to underlie the technical arguments.

Three geometric entities are considered in eqns (4) and (5) for the elasticity problem.

**Displacement and traction representation.** The real functions $N_n^{o_e} \equiv N_n^{o_e}(\xi)$ and $N_\ell^{o_e} \equiv N_\ell^{o_e}(\xi)$ of the real, natural variable $\xi \in [0,1]$ interpolate displacements and tractions, respectively, along a generic boundary segment $\Gamma_{seg} \equiv \Gamma_{seg}(\xi)$, $\xi \in [0,1]$, of the problem's whole boundary $\Gamma$. This is carried out in the frame of a *consistent* formulation, to which the reader is referred in order to fully understand our developments (we maintain that the developments in the textbooks on the BEM are not consistent). The functions $N_n^{o_e}$ interpolate displacements from *nodal* displacements $d_n$, whereas $N_\ell^{o_e}$ interpolate from traction parameters attached to boundary *loci* (attention to Ansatz 2 and eqn (20) of Dumont [6]!). Observe that *nodes* $n$ and *loci* $\ell$ are different geometric entities that may be differently allocated along the boundary.

The superscript $o_e$ in both $N_n^{o_e}$ and $N_\ell^{o_e}$ is the interpolation order in terms of Lagrangian polynomials, as we have considered in the previous papers and are considering here. As outlined in Dumont [6], $o_e = 0$ corresponds to the *constant* element and, although implicit in our developments, should be avoided for referring to a superparametric formulation (just use $o_e = 1$ for better results with the same computational effort). Our codes are implemented for the four cases $o_e = 1, 2, 3, 4$, but higher-order elements are seamlessly supported.

**Isoparametric boundary geometry description.** We also recognize in eqns (4) and (5) the boundary geometry description given by the complex $z \equiv z(\xi) - z_s = x(\xi) - x_s + i\,(y(\xi) - y_s)$, then referred to a *source* point $s$ that may be infinitesimally close to but is conceptually not on the boundary [6]. In an isoparametric formulation, the boundary Cartesian coordinates $(x, y)$ are interpolated along each boundary segment $\Gamma_{seg}$ in terms of interpolation functions $N_m^{o_e}$ ($m$ refers to key geometric points) that are linear combinations of the displacement interpolation functions $N_n^{o_e}$. However, it may not be the most advisable strategy, in general, as illustrated for the circular hole and the cut-out test of Section 2, for their circular and sinusoidal boundary patches.

**Schematic illustration.** We illustrate on the left in Fig. 3 – as already advanced in Dumont [13] – the case of two consecutive cubic ($o_e = 3$) boundary elements of a 2D elasticity problem, with $n_m = n_n = 4$ *nodes* for geometry ($\circ$) and displacements ($\odot$), which in this case coincide (as it usually occurs in an isoparametric formulation), and $n_\ell = n_n = 4$ *loci* ($\times$) for tractions, which are not at the element extremities but at distances $\epsilon \to 0$ (we do not say 'discontinuous', which is just a misconception [6]). The *points* ($*$) are for the collocation of the sources $s$ in the domain but at distances $\to 0$ from the nodal points $n$, in the frame of the CBEM. There are $n^d = n^{el}(n_n - 1) = 3n^{el}$ nodes for a total of $n^{el}$ elements that comprise the complete problem we are simulating with cubic elements. For an elasticity problem implemented in terms of real variables, the double-layer potential matrix **H** is square of order $2n^d = 2n^{el}(n_n - 1) = 6n^{el}$, and the single-layer potential matrix **G** has the same number of rows but $2n^t = 2n^{el}n_\ell = 8n^{el}$ columns, where $n^t$ is the total number of traction *loci*.

The quest 'isoparametric versus geometry-preserving' boundary description is this paper's core. However, we should first briefly address a related subject, itemized as # 1 in Section 1, which may look promising but is not.

### 4.1 Attempt to optimize *node* $n$ and *locus* $\ell$ location inside an isoparametric element

The case on the right in Fig. 3 is almost similar to the previous description, also with $n_m = n_n = 4$ *nodes* for geometry and displacements, but whose locations only coincide at the extremities, since we are now considering the abscissas of a Radau–Lobatto quadrature for

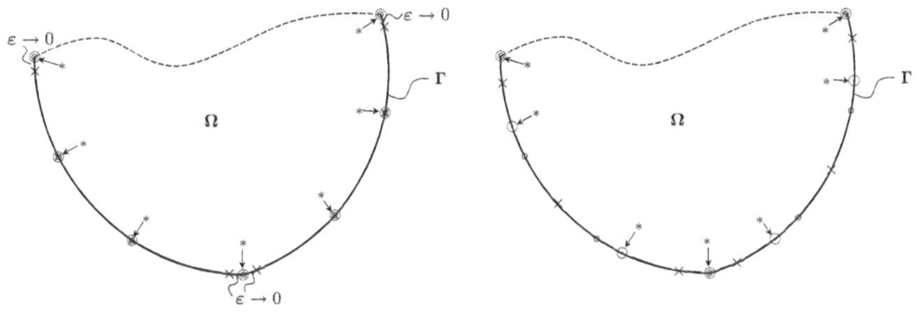


Figure 3: Two consecutive cubic elements for a 2D problem, illustrated on the left for $n_m = n_n = n_\ell = 4$ *nodes* and *loci* per element, and on the right in an optimization attempt for $n_m = n_n = 4$ *nodes* and $n_\ell = 3$ *loci*.

the displacement nodes ($\odot$) while keeping the geometry nodes ($\circ$) unaltered. Most important, we have $n_\ell = n_n - 1 = 3$ parameter *loci* ($\times$) for traction along an element at abscissas given by the roots of a Legendre polynomial, which are at finite distances from the element extremities: such implementation still satisfies the convergence Theorem 1 [6]. There are in this case $n^d = n^t = n^{el}(n_n - 1) = 3n^{el}$ *nodes* and *loci* for a total of $n^{el}$ elements. In terms of real variables, both matrices **G** and **H** are square of order $2n^d = 2n^{el}(n_n - 1) = 6n^{el}$.

The reason for such an implementation would be to improve the representation capacity of interpolation functions $N_n^{o_e}$ and $N_\ell^{o_e}$, as $n_n$ Radau–Lobatto and $n_\ell$ Legendre points lead to the accurate integral representation of polynomials of order $2n_n - 3$ and $2n_\ell - 1$, respectively. In the illustrative case of a cubic (order 3) element, we have $2 \times 4 - 3 = 5$ and $2 \times 3 - 1 = 5$, thus 5th (and not just 3rd) order polynomial representations for displacements and tractions.

Such an idea of polynomial optimization seems tempting and may deserve some numerical experimentations (we already have the code implementation). However, there are some issues to consider. First, the polynomials $N_\ell^{o_e}$ and $N_n^{o_e}$ do not feature alone in a boundary element implementation, but rather multiplied with some functions, as given in eqns (4) and (5). Then, the integral representation capacity referred to in the above paragraph – and the basis of the Gauss–Legendre quadrature – does not take place in the applications of interest. A second, not strong reasoning would be in terms of result interpretation, as the $n_n$ and $n_\ell$ locations are not as simply distributed as in the scheme on the left in Fig. 3. There is, however, a very strong argument against such an optimization attempt, which is related to the lack of smoothness in the distribution of $n_n$ and $n_\ell$ locations along a boundary patch – particularly in the frame of an adaptive mesh refinement, as we address next.

## 5 THE CONCEPT OF GEOMETRY-PRESERVING BOUNDARY PATCHES $\Gamma_{patch}$

The geometry and all relevant mesh-generation data of the circular hole of Fig. 1 are given in terms of just two semicircles and three lines of data. As given in Table 1 of Dumont [9] and to be observed in Fig. 2, the features of this topologically challenging model require 18 lines of data for complete generation of all 15 boundary patches. (We need an extra line with a dummy node to close a subboundary). The concept of *boundary patch* ($\Gamma_{patch}$) leads to a simpler and faster code for the same data entry, as an advancement of the codes reported in our recent publications. In Table 1 of Dumont [9], for instance, the first patch starts at node 1, coordinates $(0, 0)$, and finishes at node 17, coordinates $(7.2, 14)$, with the geometry deviating from the chord $1 - 17$ by the curve $y_{patch}(\zeta) = 5\sin(2\zeta\pi), \zeta \in [0, 1]$. This patch generation

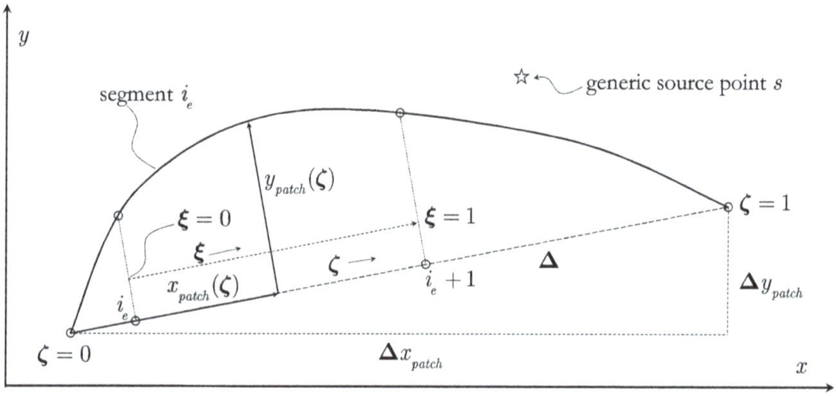


Figure 4: Generic boundary patch $\Gamma_{patch}(\zeta)$, $\zeta \in [0, 1]$ with a segment $\Gamma_{seg}(\xi)$, $\xi \in [0, 1]$.

is for quartic elements ($o_e = 4$) and 4 elements, which means that the key node number 17 is itself generated ($= 1 + o_e \times 4$). We enter for this patch that the distance from node to node varies at a geometric rate $f_{node} = 1.0$, is then constant (increasing and decreasing node distances occur for the patches connecting 17–25 and 61–65).

## 5.1 Homothetic node and element generation

Fig. 4 represents a general patch that spans from $(x_{init}, y_{init})$ to $(x_{final}, y_{final})$ and has the shape $(x_{patch}(\zeta) = \Delta\zeta, y_{patch}(\zeta))$, $\zeta \in [0, 1]$ where $y_{patch}(\zeta)$ must be entered.

We may have any topologically consistent boundary shape $y_{patch}(\zeta), \zeta \in [0, 1]$, in which 'consistent' means taking care that the Jacobian of the coordinate transformation keeps positive not only along the curved boundary but also in the complex vicinity, such as for the indicated source point $s$, for the analytical corrections of eventual quasi singularities [6]–[9], [13] to take place. Once the patch-referred coordinates are obtained, the global coordinates become, here expressed in terms of the complex $z(\zeta) = x(\zeta) + iy(\zeta)$, for simplicity,

$$z(\zeta) = z_{init} + z_{patch}(\zeta)\frac{\Delta z_{patch}}{\Delta} \tag{6}$$

In our previous codes, the outer-most loop spans all boundary elements (segments $\Gamma_{seg}$), as shown in Algorithm 1 of Dumont [6], for (computationally more involved) geometric data should be evaluated first. We are coherently proposing that the outer-most loop runs for the boundary patches, $i_{patch} = 1 \ldots n_{patch}$, with geometry pre-evaluations carried out and stored for a typical element of the patch, as shown next (see also [11]), which includes adaptive mesh refinement along a patch within the concept of *homothetic elements*. (We might have element orders $o_e$ differing from patch to patch.) Only then we run a loop for the source points $s$, evaluate the patch-related complex distance $\zeta_s$, and then the loop for all elements inside the patch, again, for which the pre-evaluations have been done.

Fig. 5 is the schematic illustration of how homothetic elements are generated, with four meshes going from node 1 through node 13, for the generation of 12 linear, 6 quadratic, 4 cubic or 3 quartic elements including eventual internal natural points $\xi_j, j = 2 \ldots o_e$, as we always have $\xi_1 = 0$ and $\xi_{o_e+1} = 1$.

In this figure, the nodes 2 to 12 are generated in such a way that the distance between consecutive nodes increases at a geometric rate $f_{node} = 1.25$. In our homothetic element

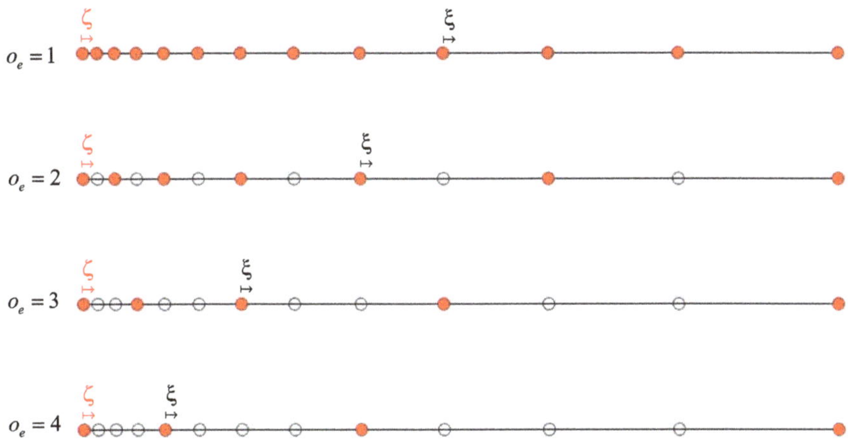


Figure 5: Schematic representation of a boundary patch (natural coordinate $\zeta \in [0,1]$) with 13 nodes (circles), whose consecutive distances grow at the rate $f_{node} = 1.25$. The homothetic subdivision for linear through quartic elements (natural coordinates $\xi_j \in [0,1]$, solid circles for the respective end nodes) is also indicated.

generation, the relative size $f_{el}$ of consecutive elements also increases exponentially:

$$f_{el} = f_{node}^{o_e} \tag{7}$$

This means that the node locations inside any element of the patch have the same representation in terms of the element natural coordinate $\xi \in [0,1]$, which is illustratively shown for some elements in the figure. This is the reason of calling 'homothetic' such combined element and node generation. Given a boundary patch $i_{patch}$, we first carry out all necessary geometric and singularity-related pre-evaluations for a representative element of the patch and only then proceed with the algorithm that takes the source points into account. These pre-evaluations include the analytical expression and storage of $N_n^{o_e}$ and $N_\ell^{o_e}$ as well as of the integrals required in the quasi-singularity corrections – always taking into account that the natural node coordinates $\xi_j, j = 1 \ldots o_e + 1$ are not equally spaced inside the element but rather reflect the distance amplification illustrated in Fig. 5. This means, for instance, that the Jacobian for coordinate transformations along straight and circular patches is constant, which leads to smaller quadrature errors. (See the second paragraph of Section 5.1.)

Let $n_\zeta$ be the numbering difference between the first and last nodes of a patch ($n_\zeta = 13 - 1 = 12$ in Fig. 5). We set

$$\tilde{\zeta} = 1 \Big/ \sum_{j=0}^{n_\zeta - 1} f_{node}^j \,, \quad \delta = 0 \tag{8}$$

and carry out the simple algorithm for the evaluation of the local $\zeta_i$ coordinates of the generated nodes in the patch:

$$\begin{aligned} &\text{for } i \text{ from } 1 \text{ to } n_\zeta + 1 \text{ do} \\ &\quad \zeta_i = \delta \tilde{\zeta} \\ &\quad \delta \leftarrow \delta f_{node} + 1 \\ &\text{end do} \end{aligned} \tag{9}$$

The coordinate transformation between patch coordinate $\zeta$ and element coordinate $\xi$ is given for the i$_th$ element, according to Fig. 5, as

$$\zeta = \zeta_i + \xi\,(\zeta_{i+1} - \zeta_i) \quad \Leftrightarrow \quad \xi = (\zeta - \zeta_i)/(\zeta_{i+1} - \zeta_i), \quad \zeta \in [\zeta_i, \zeta_{i+1}] \tag{10}$$

and the Jacobian of the coordinate transformation is

$$|J(\xi)|_{seg} = |J(\zeta)|_{patch} \left.\frac{\partial \zeta}{\partial \xi}\right|_{seg} = |J(\zeta)|_{patch}\,(\zeta_{i+1} - \zeta_i)\,, \quad \zeta \in [\zeta_i, \zeta_{i+1}] \tag{11}$$

with, in complex coordinates, $|J(\zeta)| \equiv |\partial z(\zeta)/\partial \zeta|$.

### 5.2 Evaluation of the complex $\zeta_s$ coordinate of a quasi-singular point source $z_s = x_s + iy_s$

Fig. 4 depicts a *star* for a generic source point. Such points are also indicated as *crosses* $(+)$ in Figs 1 and 2, and may be extremely close to the boundary, as illustrated in Section 2 with reference to Dumont [6]–[9], [13]. In these papers, the complex natural coordinate $\xi_s$ corresponding to a close source point $z_s = x_s + iy_s$ is evaluated iteratively for every boundary segment $\Gamma_{seg}$, as its geometry is given – in the frame of an isoparametric formulation – piecewise in terms of the interpolation functions $N_n^{o_e}(\xi)$ with local support $\xi \in [0, 1]$. In the present geometry-preserving formulation, we have a unique analytical function $z(\zeta)$ spanning a whole boundary patch, according to eqn (6). The search is then for the patch-related complex natural coordinate $\zeta_s$ in terms of the same Newton–Raphson algorithm outlined in Appendix A of Dumont [6]. Once obtained for a given patch $\Gamma_{patch}(\zeta)$, $\zeta_s$ is successively transformed into the complex $\xi_s$ for each one of the boundary segments $\Gamma_{seg}(\xi)$ according to the same eqn (10):

$$\xi_s = (\zeta_s - \zeta_i)/(\zeta_{i+1} - \zeta_i), \quad \text{for the i\_}th \text{ boundary segment} \tag{12}$$

where both $\zeta_s$ and $\xi_s$ are in general complex. When the boundary patch is either a straight segment or an arc of circle, as in Fig. 1 or in 14 of the 15 patches of Fig. 2, the natural coordinates $\zeta$ and $\xi$ of the general Fig. 4 are defined as following along the curved patch, leading to a constant Jacobian even for adaptive mesh refinement. In such particular cases, the complex source point $\zeta_s$ may be obtained analytically, which speeds up calculations [11].

### 5.3 Spaces **W**, **R** of inadmissible displacements and tractions

**Rigid-body displacements.** We have proposed at the very beginning of our developments on boundary element methods [15] a matrix **W** as the basis of rigid-body displacements in a finite domain, as expressed for elasticity. This is shown in detail in Dumont [6], where it is set in eqn (18) of **Definition 1** that the rigid-body displacements $u_{ik}^r$ along the boundary are linear combinations of the displacements $N_n^{o_e}$. However, this only holds in the isoparametric formulation. In the present geometry-preserving context for 2D problems, the three rigid-body displacements (two translations and one rotation) at a given point of the boundary are

$$\begin{bmatrix} u_1^r & u_2^r & u_3^r \\ \bar{u}_1^r & \bar{u}_2^r & \bar{u}_3^r \end{bmatrix} \Leftarrow \begin{bmatrix} 1 & i & iz(\zeta) \\ 1 & -i & -i\bar{z}(\zeta) \end{bmatrix} \quad \text{at a point of the boundary patch} \tag{13}$$

here defined in terms of the complex variable, so that we only evaluate the first matrix row.

**Inadmissible tractions.** Building up on a 1998 paper [3], it is shown in Dumont [6] – and explored subsequently also in the context of complex variables [9], [13] – that a *consistent* boundary element formulation requires that, for a finite body, if the rigid-body amount of displacements $(\mathbf{d}-\mathbf{d}^p)$ in eqn (3) cannot be transformed into forces, conversely, the amount of non-equilibrated tractions $(\mathbf{t}-\mathbf{t}^p)$ cannot be transformed into displacements:

$$\mathbf{HW}=\mathbf{0} \quad \Leftrightarrow \quad \mathbf{G}_{ad}\mathbf{R}=\mathbf{0} \tag{14}$$

This reasoning leads to the evaluation of the rigid-body displacement amount in the expression of a fundamental solution, embedded in the single-layer potential matrix $\mathbf{G}_{ad}$, where the subscript means the filtered, *admissible* part of the matrix. The reader is referred to Dumont [6], where the matter is illustrated didactically. The formulation in terms of a complex variable is shown in Appendix A.2.2 of Dumont [9]. In the case of a geometry-preserving formulation,

$$R_{\ell k}=|J|_{(\text{at }\ell)}\int_0^1\begin{bmatrix}1 & i & iz\left(\zeta(\xi)\right)\\ 1 & -i & -i\bar{z}\left(\zeta(\xi)\right)\end{bmatrix}N_\ell^{o_e}(\xi)\mathrm{d}\xi \quad \text{for a boundary element} \tag{15}$$

obtained using Gauss–Legendre quadrature along each segment of the whole boundary, as $N_\ell^{o_e}$ has local support for $\xi\in[0,1]$ [6], [9], [11], [13].

The left part of eqn (14) is rather axiomatic also in the present context and has been checked numerically for the topologically challenging example of Fig. 2: errors are only due to the Gauss–Legendre quadrature of the problem's regular integrals.

## 6 CONCLUDING REMARKS

The developments proposed in Dumont [9], [10] as the complex-variable counterpart of Dumont [6]–[8] for 2D potential and elasticity problems could be further improved, as shown in this short contribution and to be thoroughly assessed in a paper in preparation [11]. Further to properly addressing the three entities – *boundary nodes n* (for potentials or displacements), *boundary loci* $\ell$ (to which normal fluxes or tractions are referred), and *domain points s*, at which we collocate the singular sources in the context of an isoparametric formulation – we show that a *geometry-preserving* concept must be explicitly resorted to if numerical precision and accuracy are deemed relevant in the numerical simulation of real-world problems. The convergence Theorem proposed in Dumont [6] is no longer applicable but this seems to be compensated by the attained geometric description of a physical model – particularly when dealing with topologically challenging configurations.

We use the concept of *geometry-preserving boundary patches* $\Gamma_{patch}$, along which boundary nodes and elements are adaptively refined in a *homothetic* concept. This avoids distortions in the problem's geometry description and leads to more robust simulations and more accurate and reliable results, as assessed in Dumont [11]. Given a source point $(x_s, y_s)$, we only need to evaluate its complex location $\zeta_s$ once for a whole boundary patch $\Gamma_{patch}$. This and the simplifications related to the *homothetic* mesh refinement, particularly using the complex $z=x+iy$, lead to the assembling time of all relevant matrices significantly smaller than in the original code. The analytical, correction terms to be accrued for quasi-singularities are the same ones previously proposed.

## ACKNOWLEDGEMENTS

This project was supported by the Brazilian federal agencies CAPES and CNPq, as well as by the state agency FAPERJ.

## REFERENCES

[1] Dumont, N.A., On the efficient numerical evaluation of integrals with complex singularity poles. *Engineering Analysis with Boundary Elements*, **13**, pp. 155–168, 1994.

[2] Dumont, N.A. & Noronha, M., A simple, accurate scheme for the numerical evaluation of integrals with complex singularity poles. *Computational Mechanics*, **22**(1), pp. 42–49, 1998.

[3] Dumont, N.A., An assessment of the spectral properties of the matrix G used in the boundary element methods. *Computational Mechanics*, **22**(1), pp. 32–41, 1998.

[4] Dumont, N.A., The boundary element method revisited. *WIT Transactions on Modelling and Simulation*, vol. 50, WIT Press: Southampton and Boston, pp. 227–238, 2010.

[5] Dumont, N.A., The collocation boundary element method revisited: Perfect code for 2D problems. *International Journal of Computational Methods and Experimental Measurements*, **6**(6), pp. 965–975, 2018.

[6] Dumont, N.A., The consistent boundary element method for potential and elasticity: Part I – Formulation and convergence theorem. *EABE – Engineering Analysis with Boundary Element Methods*, **149**, pp. 127–142, 2023.

[7] Dumont, N.A., The consistent boundary element method for potential and elasticity: Part II – Machine-precision numerical evaluations for 2D problems. *EABE – Engineering Analysis with Boundary Element Methods*, **149**, pp. 92–111, 2023.

[8] Dumont, N.A., The consistent boundary element method for potential and elasticity: Part III – Topologically challenging numerical assessments for 2D problems. *EABE – Engineering Analysis with Boundary Element Methods*, **151**, pp. 548–564, 2023.

[9] Dumont, N.A., Complex-variable, high-precision formulation of the consistent boundary element method for 2D potential and elasticity problems. *EABE – Engineering Analysis with Boundary Element Methods*, **152**, pp. 552–574, 2023.

[10] Dumont, N.A., Real- and complex-variable implementations of the consistent boundary element method in two-dimensional elasticity: A comparative assessment. *WIT Transactions on Engineering Sciences*, vol. 135, WIT Press: Southampton and Boston, pp. 55–66, 2023.

[11] Dumont, N.A., Consistent boundary element method for two-dimensional problems with geometry-preserving, homothetic element generation. *EABE – Engineering Analysis with Boundary Element Methods*, 2024. To be submitted.

[12] Cottrell, J.A., Hughes, T.J.R. & Bazilevs, Y., *Isogeometric Analysis: Toward Integration of CAD and FEA*. John Wiley and Sons, 2009.

[13] Dumont, N.A., Consistency, precision, and accuracy assessment of the collocation boundary element method for two-dimensional problems of potential and elasticity. *Archive of Applied Mechanics*, **9**, 2024.

[14] Sadd, M.H., *Elasticity, Theory, Applications, and Numerics*, Elsevier, 2005.

[15] Dumont, N.A., The hybrid boundary element method: an alliance between mechanical consistency and simplicity. *Applied Mechanics Reviews*, **42**(11), pp. S54–S63, 1989.

# SOLUTION OF CORROSION PROBLEMS USING THE BOUNDARY ELEMENT METHOD

NOUF R. ALTALHI[1,2], DANIEL LESNIC[1] & STEPHEN D. GRIFFITHS[1]
[1]Department of Applied Mathematics, University of Leeds, UK
[2]Department of Mathematics, University of Taif, Saudi Arabia

## ABSTRACT

In the practical setting of electrical impedance tomography, a current flux is injected through an electrode attached to the boundary and the boundary electrical potential is measured. This is numerically simulated by solving using the boundary element method (BEM) a direct problem for the Laplace equation with a nonlinear Butler–Volmer boundary condition over the boundary of a buried pipe that takes into account the principles of chemical corrosion. Our methodology is placed within the mathematical context of the Cauchy problem for the Laplace equation in a rectangular domain enclosing the corroded pipe whose boundary is inaccessible to prescribing any conditions on it. Given the Cauchy data, i.e. the specification of both the flux and potential, on an accessible portion of the boundary of the solution domain, the task is to determine the unknown potential and the flux on the inaccessible boundary of the pipe and the potential everywhere else inside the solution domain. This approach holds the promise of providing a non-invasive, accurate and effective means of detecting internal pipe corrosion, thus contributing significantly to maintaining safety standards in nuclear power plants. Since the inverse problem is ill-posed the resulting system of equations is ill-conditioned and the truncated singular value decomposition is employed in order to obtain stable and accurate numerical solutions. Other related inverse problems arising when modelling corrosion are also discussed.
*Keywords: inverse problem, boundary elements, corrosion.*

## 1 INTRODUCTION

Offshore structures such as sea platforms, bridges or ships made of metal are subjected to destructive corrosion attacks through electrochemical reactions with the air/water surroundings containing oxygen, or oxidizing agents, that affect their integrity [1]. Ensuring the integrity of piping systems is critical to management of nuclear power plants, where internal corrosion poses a significant risk to operational stability and safety. In order to understand the corrosion of a structure one needs to analyse the data on the electrical potential and the current density distribution over its boundary.

## 2 MATHEMATICAL FORMULATION

We consider an ion solution (electrolyte) occupying a bounded medium $\Omega$ within which the continuity equation

$$\nabla \cdot \underline{J} = 0 \quad \text{in} \quad \Omega \tag{1}$$

holds, where the current density vector $\underline{J}$ is related to the electrical potential $\phi$ through

$$\underline{J} = -\sigma \nabla \phi, \tag{2}$$

where $\sigma$ is the conductivity of the medium, which may be isotropic or an-isotropic. In (1) we have assumed, for simplicity, that any source (or sink) generating current was absent. Introduction of (2) into (1) leads to the elliptic equation

$$-\nabla \cdot (\sigma \nabla \phi) = 0 \quad \text{in} \quad \Omega. \tag{3}$$

WIT Transactions on Engineering Sciences, Vol 136, © 2024 WIT Press
www.witpress.com, ISSN 1743-3533 (on-line)
doi:10.2495/BE470021

In this study, we consider only steady-state corrosion, with transient phenomena modelled by the parabolic heat equation being deferred to future study.

For simplicity, we assume that the electrolyte solution is homogeneous, i.e. $\sigma =$ constant, such that eqn (3) simplifies into the Laplace equation

$$\nabla^2 \phi = 0 \quad \text{in} \quad \Omega. \tag{4}$$

Nonlinear electrolytes with conductivity $\sigma = \sigma(\phi)$ dependent on the potential $\phi$ may also be considered and reduced to the Laplace equation via the Kirchhoff transformation $\psi = \int \sigma(\phi) d\phi$.

As for the boundary conditions associated to (4) we consider that a part of the boundary is electro-chemically active, where the oxidation-reduction reactions take place, while the remaining part is inactive, although we may inject a current flux on it. On the electro-chemically active metal-electrolyte boundary, owing to the polarization effect, the current flux is a nonlinear function of the potential difference $\eta_s := \phi - \phi_{eq}$ of the form [1],

$$J_e := \underline{J} \cdot \underline{n} = J_0 f(\phi - \phi_{eq}), \tag{5}$$

where $\underline{n}$ is the outward unit normal to the electrode's boundary, $\phi_{eq}$ is the equilibrium potential (also called the Nernst potential) of the electrode, $J_0$ is the exchange current density and $f$ is a nonlinear polarization function.

If $|\eta_s|$ is small, then the nonlinear function in (5) is given by the Butler–Volmer equation [2],

$$f(\eta_s) = \exp(\beta\gamma\eta_s) - \exp(-(1-\beta)\gamma\eta_s), \tag{6}$$

where $\beta \in (0,1)$ is a symmetry kinetic parameter called the transfer coefficient, $\gamma = F/(RT)$, where $F$ is the Faraday's constant, $R$ is the universal gas constant, $T$ is the absolute temperature and the charge number of the dissolved or deposited species was taken to be equal to unity, for simplicity.

Based on (6), the boundary condition (5) can be written as [3],

$$-\sigma\frac{\partial \phi}{\partial n} = \underline{J} \cdot \underline{n} = J_0\big(\exp(\beta\gamma\eta_s) - \exp(-(1-\beta)\gamma\eta_s)\big). \tag{7}$$

Defining

$$u := \gamma\eta_s = \frac{(\phi - \phi_{eq})F}{RT}, \tag{8}$$

we obtain from (4) and (7) that

$$\nabla^2 u = 0 \quad \text{in} \quad \Omega \tag{9}$$

and

$$\frac{\partial u}{\partial n} = \lambda\big(\exp(\beta u) - \exp(-(1-\beta)u)\big) \quad \text{on} \quad \Gamma_0, \tag{10}$$

where $\Gamma_0$ denotes the corroded boundary and

$$\lambda = -\frac{\gamma J_0}{\sigma} = -\frac{F J_0}{\sigma RT}. \tag{11}$$

In general, $\lambda < 0$ on the active region where $|\eta_s|$ is small (at the anodic or cathodic surface) and $\lambda > 0$ in the transition region elsewhere [4].

The remaining part of the boundary $\partial\Omega\backslash\Gamma_0$ is assumed electrochemically inactive and we prescribe a Neumann condition

$$\frac{\partial u}{\partial n} = g \quad \text{on} \quad \partial\Omega\backslash\Gamma_0, \tag{12}$$

where $g$ represents an imposed current flux.

## 3 MATHEMATICAL ANALYSIS

We write again the mathematical model of corrosion given by

$$\begin{cases} \nabla^2 u = 0 & \text{in } \Omega, & (13) \\ \dfrac{\partial u}{\partial n} = \lambda f(u) + g & \text{on } \Gamma_0, & (14) \\ \dfrac{\partial u}{\partial n} = g & \text{on } \partial\Omega \setminus \Gamma_0, & (15) \end{cases}$$

where possibly $g|_{\Gamma_0} = 0$ and

$$f(u) = \exp(\beta u) - \exp(-(1-\beta)u), \tag{16}$$

where $\beta \in (0,1)$. Given $g \in L^2(\partial\Omega)$, the weak solution $u \in H^1(\Omega)$ of the problem (13)–(15) satisfies

$$\int_\Omega \nabla u \cdot \nabla v \, d\Omega = \int_{\Gamma_0} \lambda f(u)\, v \, d\Gamma_0 + \int_{\partial\Omega} g \, v \, ds, \quad \forall v \in H^1(\Omega). \tag{17}$$

Then the following theorem established in Vogelius and Xu [4] gives results on the unique solvability of the direct problem (13)–(15).

**Theorem 3.1.** *Given $g \in L^2(\partial\Omega)$, there exist $\lambda_0 > 0$ such that the solution to the problem (13)–(15) exists for any $\lambda \in (-\infty, \lambda_0)$. Moreover, for $\lambda \in (-\infty, 0]$ this solution is unique (except when $\lambda = 0$, which is unique up to a constant).*

*When $\Omega$ is a disk in two-dimensions, the numerical investigations performed in Bryan and Vogelius [5] pointed out that there are infinitely many solutions to (13)–(16) for any $\lambda > 0$.*

The Maclaurin series expansion

$$f(u) = \exp(\beta u) - \exp(-(1-\beta)u) \approx u + \frac{2\beta - 1}{2} u^2 + ..., \tag{18}$$

results in the linearized problem (near $|u| \ll 1$) given by

$$\begin{cases} \nabla^2 u = 0 & \text{in } \Omega, & (19) \\ \dfrac{\partial u}{\partial n} = \lambda u + g & \text{on } \Gamma_0, & (20) \\ \dfrac{\partial u}{\partial n} = g & \text{on } \partial\Omega \setminus \Gamma_0, & (21) \end{cases}$$

and the weak formulation (17) becomes

$$\int_{\Omega} \nabla u \cdot \nabla v \, d\Omega = \int_{\Gamma_0} \lambda u \, v \, d\Gamma_0 + \int_{\partial\Omega} g \, v \, ds, \quad \forall v \in H'(\Omega). \tag{22}$$

Then the unique solvability of the linear direct problem (19)–(21) holds for any real values of $\lambda$ outside a countable set $\{\lambda_n\}_{n\geq 1}$ of positive integers ordered as $0 < \lambda_1 < \lambda_2 < \cdots$, see Vogelius and Xu [4].

## 4 THE BOUNDARY ELEMENT METHOD (BEM)

The BEM is well-suited for approaching numerically the direct problem (13)–(15) since the governing elliptic Laplace's equation (13) possesses an explicit fundamental solution given by

$$G(\underline{p};\underline{p}') = \begin{cases} -\frac{1}{2\pi} \ln |\underline{p} - \underline{p}'| & \text{in two-dimensions} \\ \frac{1}{4\pi|\underline{p}-\underline{p}'|} & \text{in three-dimensions.} \end{cases} \tag{23}$$

Then classical application of Green's formula and imposition of the boundary conditions (14) and (15) results in the boundary integral representation of the solution as

$$\eta(\underline{p})u(\underline{p}) = \int_{\Gamma_0} \lambda(\underline{p}') f(u(\underline{p}')) \, G(\underline{p};\underline{p}') d\Gamma_0 + \int_{\partial\Omega} g(\underline{p}') \, G(\underline{p};\underline{p}') ds \\ - \int_{\partial\Omega} u(\underline{p}') \frac{\partial G}{\partial n(\underline{p}')}(\underline{p};\underline{p}') ds, \quad \underline{p} \in \overline{\Omega} = \Omega \cup \partial\Omega, \tag{24}$$

where $\eta(\underline{p}) = 1$ if $\underline{p} \in \Omega$ and $\eta(\underline{p}) = 0.5$ if $\underline{p}$ is on a smooth part of $\partial\Omega$. For simplicity, let us describe the BEM in two-dimensions only. Using a constant BEM approximation the function $u$ is assumed to be piecewise constant and take the value at the midpoint (node) $\tilde{\underline{p}}_j = (\underline{p}_j + \underline{p}_{j-1})/2$ of each straight-line boundary element $S_j = [\underline{p}_{j-1}, \underline{p}_j]$ for $j = \overline{1, M}$, discretising in an anti-clockwise sense the boundary $\partial\Omega$ where $M$ denotes the number of boundary elements and $\underline{p}_M = \underline{p}_0$ by convention, namely,

$$u(\underline{p}') = u(\tilde{\underline{p}}_j) =: u_j, \quad \forall \underline{p}' \in S_j, \quad j = \overline{1, M}. \tag{25}$$

We also approximate

$$g(\underline{p}') = g(\tilde{\underline{p}}_j) =: g_j, \quad \forall \underline{p}' \in S_j, \quad j = \overline{1, M}, \tag{26}$$

$$\lambda(\underline{p}') = \lambda(\tilde{\underline{p}}_j) =: \lambda_j, \quad \forall \underline{p}' \in S_j, \quad j = \overline{1, N}. \tag{27}$$

Assume that $\Gamma_0$ is discretised by the first $N$ boundary elements. Then, based on (25)–(27), eqn (24) yields the BEM approximation

$$\eta(\underline{p})u(\underline{p}) = \sum_{j=1}^{N} \lambda_j f(u_j) \int_{S_j} G(\underline{p};\underline{p}') dS_j + \sum_{j=1}^{M} g_j \int_{S_j} G(\underline{p};\underline{p}') dS_j \\ - \sum_{j=1}^{M} u_j \int_{S_j} \frac{\partial G}{\partial n(\underline{p}')}(\underline{p};\underline{p}') dS_j, \quad \underline{p} \in \overline{\Omega}. \tag{28}$$

Denoting by

$$A_j(\underline{p}) = \int_{S_j} G(\underline{p};\underline{p}')dS_j, \quad B_j(\underline{p}) = \int_{S_j} \frac{\partial G}{\partial n}(\underline{p};\underline{p}')dS_j, \quad j=\overline{1,M}, \tag{29}$$

which can be evaluated analytically based on the geometry of the triangle $\Delta\, \underline{p}\, \underline{p}_{j-1}\, \underline{p}_j$, see e.g. Jaswon and Symm [6], the boundary integral equation (28) can be rewritten as

$$\eta(\underline{p})u(\underline{p}) = \sum_{j=1}^{N} A_j(\underline{p})\lambda_j f(\underline{u}_j) + \sum_{j=1}^{M} A_j(\underline{p})g_j - \sum_{j=1}^{M} B_j(\underline{p})u_j, \qquad \underline{p} \in \overline{\Omega}, \tag{30}$$

where we are assuming that $\Gamma_0$ is discretised by the first $N$ boundary elements.

Collocating (30) at the boundary nodes $(\tilde{\underline{p}}_i)_{i=\overline{1,M}}$ results in the following system of $M$ nonlinear equations with $M$ unknowns $(u_j)_{j=\overline{1,M}}$:

$$\sum_{j=1}^{N} (A_{ij}\lambda_j f(u_j) + B_{ij}u_j) + \sum_{j=N+1}^{M} B_{ij}u_j = -\sum_{j=1}^{M} A_{ij}g_j \quad \text{for} \quad i=\overline{1,M}, \tag{31}$$

where

$$A_{ij} = A_j(\tilde{\underline{p}}_i) \ \text{ and } \ B_{ij} = -B_j(\tilde{\underline{p}}_i) - \frac{1}{2}\delta_{ij},$$

and $\delta_{ij}$ is the Kronecker-delta tensor.

For the linear boundary condition $\partial_n u = \lambda u$ on $\Gamma_0$, the system (31) becomes linear and is given by

$$\sum_{j=1}^{N}(A_{ij}\lambda_j + B_{ij})u_j + \sum_{j=N+1}^{M} B_{ij}u_j = -\sum_{j=1}^{M} A_{ij}g_j \quad \text{for} \quad i=\overline{1,M}. \tag{32}$$

### 4.1 Numerical discretisation

For the two-dimensional bounded geometry of $\Omega$, we consider a doubly-connected annular domain consisting of an infinitely-long cylindrical circular buried pipe, which is subjected to a corrosion attack. Due to the symmetry of the problem, we can consider only a quadrant of the geometry, as pictured, along with the relevant boundary conditions, in Fig. 1. The boundary $\Gamma_0$ is the first quarter of the unit circle, whilst $\Gamma = \partial\Omega \setminus \Gamma_0$ is the remaining part consisting of the horizontal and vertical portions $\Gamma_i$ for $i=\overline{1,4}$. Then, due to symmetry, clearly $g=0$ on $\Gamma_1 \cup \Gamma_4$. Also, by assuming that the top wall is adiabatic we have that $g=0$ on $\Gamma_3$. On the vertical wall $\Gamma_2$, we impose a current flux over an electrode of length $2\epsilon$ centered at $(2, \frac{1}{3})$ which is drawn out over another electrode of length $2\epsilon$ centered at $(2, \frac{2}{3})$, where $\epsilon > 0$ is a small number, typically $\epsilon = 0.1$. Therefore, on $\Gamma_2$, the current flux is given by

$$g(2,y) = \begin{cases} 0 & \text{if} \quad 0 \le y < \frac{1}{3} - \epsilon, \\ \frac{1}{2\epsilon} & \text{if} \quad \frac{1}{3} - \epsilon \le y \le \frac{1}{3} + \epsilon, \\ 0 & \text{if} \quad \frac{1}{3} + \epsilon < y < \frac{2}{3} - \epsilon, \\ -\frac{1}{2\epsilon} & \text{if} \quad \frac{2}{3} - \epsilon \le y \le \frac{2}{3} + \epsilon, \\ 0 & \text{if} \quad \frac{2}{3} + \epsilon < y \le 2. \end{cases} \tag{33}$$

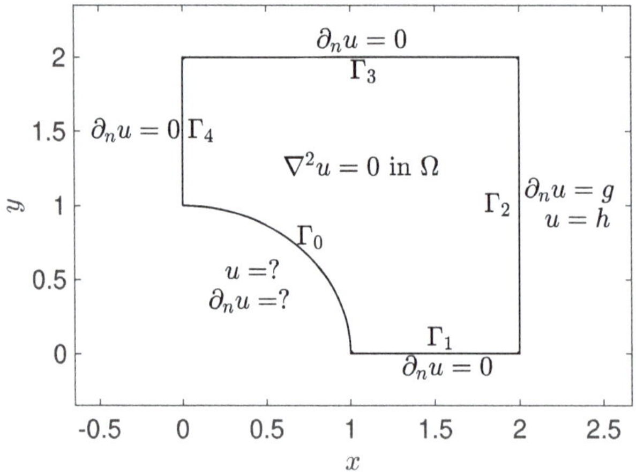


Figure 1: Cauchy inverse problem formulation.

We also take $g = 0$ on $\Gamma_0$ in which case the boundary condition (20) becomes

$$\frac{\partial u}{\partial n} = \lambda f(u) \quad \text{on} \quad \Gamma_0, \tag{34}$$

where $f(u) = u$ in the linear case and $f(u) = 2\sinh(u/2)$ in the nonlinear case.

The boundary $\partial\Omega$ is uniformly discretized in an anti-clockwise direction starting from the point (0,1), with each of the boundaries $\Gamma_0, \Gamma_1$, and $\Gamma_4$ using $N$ boundary elements, and $\Gamma_2$ and $\Gamma_3$ each using $2N$ boundary elements. This way the total number of boundary elements discretising $\partial\Omega$ is $M = 7N$.

The numerical solution of the direct problem given by eqns (13)–(15) is obtained using the BEM based on solving the system of eqns (31) (or (32)). We have also applied the method of fundamental solutions (MFS), Karageorghis and Lesnic [7], to solving the direct problem (13)–(15) and we have obtained similar results to those obtained by the BEM, re-ensuring that the numerical BEM solution is accurate.

Next, the data $u\,|_{\Gamma_2} =: h$ is employed as an extra piece of information to solve the Cauchy inverse problem depicted in Fig. 1, and given by

$$\begin{cases} \nabla^2 u = 0 & \text{in } \Omega, \quad (35) \\ \dfrac{\partial u}{\partial n} = g & \text{on } \partial\Omega \setminus \Gamma_0, \quad (36) \\ u = h & \text{on } \Gamma_2. \quad (37) \end{cases}$$

This inverse problem arises due to the inaccessibility of measuring along the corroded boundary $\Gamma_0$ of the buried pipe. Note that the inverse problem (35)–(37) is linear and ill-posed (the solution is unique but it does not depend continuously on the input data $h$), whilst the direct problem (13)–(15) is non-linear (through the boundary condition (14) if $f(u)$ is nonlinear function of $u$) and well-posed (under the conditions stated in Theorem 3.1).

The BEM numerical discretisation of the inverse problem (35)–(37) results in the following overdetermined system of $M$ linear equations in $(M-N)$ unknowns:

$$\sum_{j=1}^{N} A_{ij}u'_j + \sum_{j=1}^{2N} B_{ij}u_j + \sum_{j=4N+1}^{M} B_{ij}u_j$$
$$= - \sum_{j=N+1}^{M} A_{ij}g_j - \sum_{j=2N+1}^{4N} B_{ij}h_j \quad \text{for } i=\overline{1,M}, \tag{38}$$

where, based on the BEM with constant elements,

$$u'_j := \frac{\partial u}{\partial n}(\tilde{\underline{p}}_j) = \frac{\partial u}{\partial n}(\underline{p}'), \quad \forall \underline{p}' \in S_j, \quad j=\overline{1,N},$$
$$h_j := h(\tilde{\underline{p}}_j) = h(\underline{p}'), \quad \forall \underline{p}' \in S_j, \quad j=\overline{2N+1,4N}.$$

In a matrix form, the system of eqns (38) can be written as

$$K\underline{X} = \underline{b}, \tag{39}$$

where

$$K_{ij} = \begin{cases} A_{ij} & \text{for } j=\overline{1,N}, \\ B_{ij} & \text{for } j=\overline{1,2N} \text{ and } j=\overline{4N+1,M}, \end{cases} \quad \text{and } i=\overline{1,M},$$
$$\underline{X} = \left( (u'_j)_{j=\overline{1,N}}, (u_j)_{j=\overline{1,2N}}, (u_j)_{j=\overline{4N+1,M}} \right)^{\mathrm{T}},$$
$$b_i = - \sum_{j=N+1}^{M} A_{ij}g_j - \sum_{j=2N+1}^{4N} B_{ij}h_j \quad \text{for } i=\overline{1,M}.$$

## 5 NUMERICAL RESULTS AND DISCUSSION

For simplicity, we take $\lambda = 1$ and illustrate the numerical results obtained for the linearised boundary condition $\partial_n u = \lambda u$ on $\Gamma_0$. Although not illustrated, it is reported that similar results have been obtained for the nonlinear boundary condition $\partial_n u = 2\lambda \sinh(u/2)$ on $\Gamma_0$, due to the approximation (18) being valid in a small neighbourhood of the origin (for the direct problem with the data $g$ given by (33) on $\Gamma_0$ and zero elsewhere, we have obtained that the range of $u|_{\Gamma_0}$ is within $(-0.2, 0.2)$, see e.g. Fig. 3(a)).

We first solved the direct problem based on the BEM system of eqns (38) with $N = 30$ to obtain the data $h = u|_{\Gamma_2}$ to be used in solving the inverse problem based on the BEM system of eqns (39). Since the Cauchy problem (35)–(37) is ill-posed, the resulting system of eqns (39) is ill-conditioned. This is reflected in the behaviour of the normalised singular values of the matrix $K$ (of dimension $M \times (M-N) = 210 \times 180$) displayed in Fig. 2. The normalised singular values start off at a maximum value of 1 (since they are normalized by the largest singular value) and exhibit a steep decline, eventually spanning several orders of magnitude after about 150 singular values. This characteristic decay pattern is indicative of the presence of both significant and negligible components in the data structure. It suggests that the matrix from which these singular values are derived may be well-conditioned in the higher index range but becomes increasingly ill-conditioned as the index increases. Such a plot is crucial in the context of regularization, where one may truncate the spectrum to mitigate the effects of noise amplification, retaining only the most significant singular values for obtaining a stable solution [8].

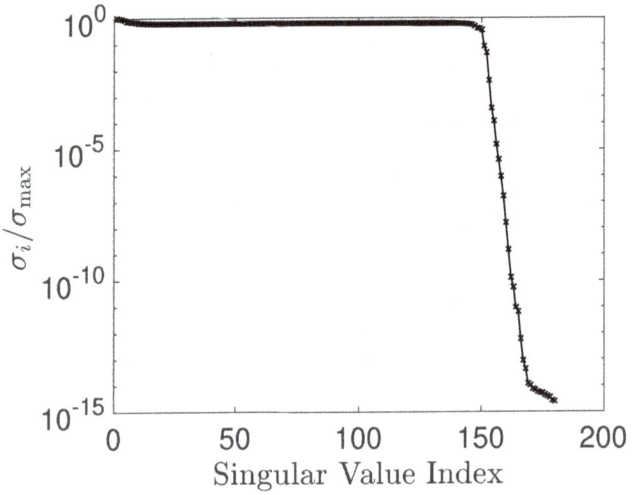


Figure 2: Normalized singular values of the matrix $K$.

Figs 3(a) and 3(b) present the numerical solutions for $u$ and $\partial_n u$ on $\Gamma_0$ obtained using the untruncated singular value decomposition method (SVD) for solving the system of eqns (39), whilst Figs 3(c) and 3(d) present the numerical solutions obtained using the truncated singular value decomposition method (TSVD) taking into account only the first 166 singular values. Employing the SVD method produced an accurate numerical solution for $u$ on $\Gamma_0$, as shown in Fig. 3(a). However, the numerical solution for the normal derivative $\frac{\partial u}{\partial n}$ on $\Gamma_0$ shown in Fig. 3(b) manifests expected instabilities. Discarding the very low singular values results in stable approximations illustrated in Fig. 3(d) by truncating the SVD expansion at 166 singular values.

To investigate further the stability of the TSVD, we invert noisy boundary temperature measured data on $\Gamma_2$, numerically simulated as

$$h_j^{noisy} = h_j(1 + p\rho_j), \quad j = \overline{(2N+1), 4N}, \tag{40}$$

where $p$ represents the percentage of multiplicative noise and $(\rho_j)_{j=\overline{(2N+1),4N}}$ are random variables drawn from a uniform distribution in $[-1, 1]$. Fig. 4 presents the numerical results obtained without and with truncation in the SVD, when $p = 1\%$ noisy data (40) are inverted. From this figure the role of appropriate truncation acting as a stable regularization can be observed. The choice of the truncation number in the TSVD depends on the amount of noise with which the data $h$ is contaminated by (40) and can be chosen according to the discrepancy principle or the L-curve method [9].

## 6 OTHER INVERSE PROBLEMS ARISING IN CORROSION

Other inverse problems related to possible unknown quantities present in eqn (14) can be formulated [10], [11], as follows.

### 6.1 Determination of the coefficient of corrosion

Another inverse problem arising in corrosion engineering requires determining the coefficient of corrosion $\lambda$ on $\Gamma_0$, see e.g. [12], [13], in case it is unknown in the boundary condition

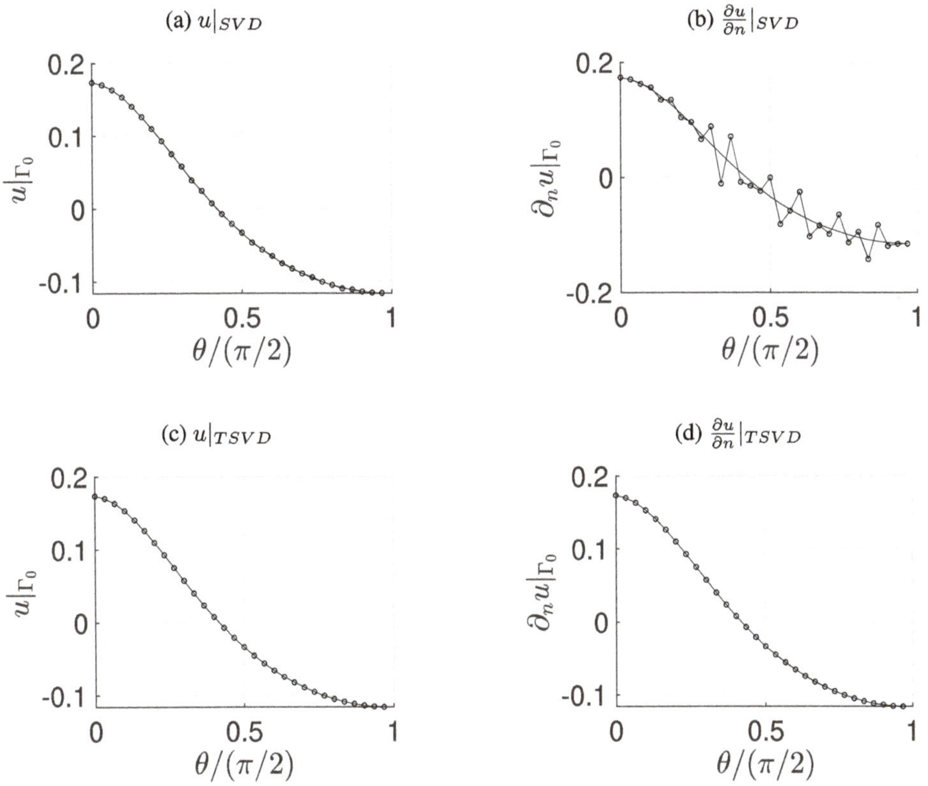


Figure 3: The numerical BEM solutions for $u$ and the normal derivative $\frac{\partial u}{\partial n}$ on $\Gamma_0$ of the inverse problem ($-\circ-$) with noiseless data obtained using: (a,b) the untruncated SVD; and (c,d) the TSVD truncated at 166 singular values, in comparison with the BEM solutions of the direct problem (——).

(14). In our setup, once the values of $u$ and $\partial_n u$ have been accurately determined on $\Gamma_0$, the coefficient of corrosion $\lambda$ can simply be estimated from (14) as

$$\lambda_j = \frac{u'_j - g_j}{f(u_j)}, \quad j = \overline{1, N}, \tag{41}$$

with the convention that the l'Hopital rule may need to be employed if a $0/0$ non-determination occurs at isolated points on $\Gamma_0$.

For the example considered in the previous section, for noiseless data, i.e. $p = 0$ in (40), employing eqn (41) yields a very accurate estimation of the exact coefficient $\lambda_{exact}(\theta) \equiv 1$ for $\theta \in [0, \pi/2]$, simply dividing the numerical solution of Fig. 3(d) by that of Fig. 3(c). No inaccuracies were obtained close to the end points $\theta = 0$ or $\theta = \pi/2$ of the circular arc $\Gamma_0$ nor close to the point $\theta \approx \frac{\pi}{2} \times 0.4$, where both $u$ and $\partial_n u$ vanish. However, in case of noisy data, by dividing the numerical solution of Fig. 4(d) by that of Fig. 4(c) results in a highly oscillatory and unbounded retrieval of the coefficient of corrosion. In this case, in order to restore stability, nonlinear iterative regularization needs to be employed in a future work.

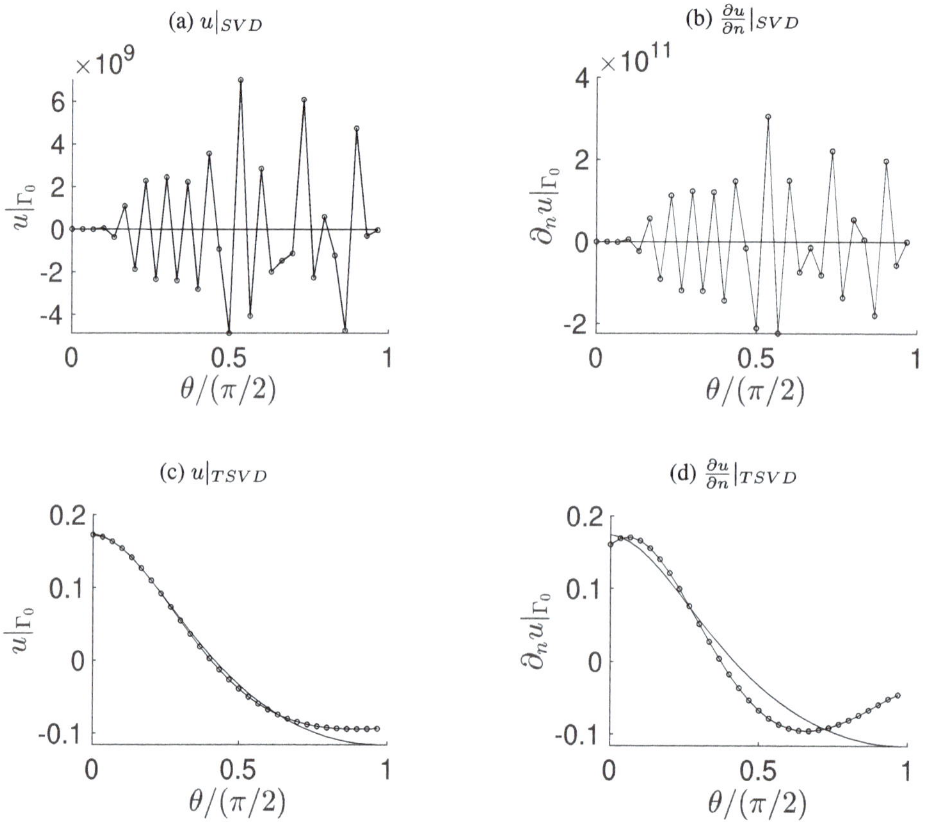


Figure 4: The numerical BEM solutions for $u$ and the normal derivative $\frac{\partial u}{\partial n}$ on $\Gamma_0$ of the inverse problem $(-\circ-)$ with $p = 1\%$ noisy data obtained using: (a,b) the untruncated SVD; and (c,d) the TSVD truncated at 152 singular values, in comparison with the BEM solutions of the direct problem (—).

### 6.2 Determination of the law of corrosion

Another inverse problem relates to the identification of the law of corrosion $f(u)$ present in the boundary condition (14), see e.g. [14], [15]. In this case, one can try to plot $u$ versus $\partial_n u$ from Figs 3(c) and 3(d), in case of noiseless data, or from Figs 4(c) and 4(d), in case of noisy data, and try to infer from the graph obtained the behaviour of the function $f(u)$, as a function of $u$. Again, as before, this simple graphical method may work in case of error-free data, but it becomes less effective in case of noisy data. In order to overcome this situation, regularization by discretisation [16] could be employed.

### 6.3 Determination of the corroded boundary

In the physical case that the corroded boundary $\Gamma_0$ is unknown, the single measurement (37) of $u$ on $\Gamma_2$ is, in general, not sufficient to determine $\Gamma_0$, see Cakoni and Kress [17]. However, two linearly independent pairs of Cauchy data $(u, \partial_n u) = (g_i, h_i)$ for $i = 1, 2$ are sufficient to determine not only the corroded boundary $\Gamma_0$ but also the coefficient of corrosion $\lambda$, in the case where the law of corrosion is linear [18].

## 7 CONCLUSIONS

This paper has presented a thorough investigation of the BEM for solving inverse problems under both noise-free and noisy input data for linear and non-linear boundary-condition models of non-destructive pipe corrosion testing. The numerically-obtained results have demonstrated that the TSVD provides a reliable stable method of solution for recovering the unknown data on the corroded boundary. The findings emphasize the necessity of methodological rigor and the need for robust computational strategies adaptable to data imperfections.

## ACKNOWLEDGEMENTS

In deep appreciation for the generous sponsorship provided by Taif University, N.R. Altalhi extends heartfelt gratitude. No data are associated with this article. For the purpose of open access, the authors have applied a Creative Commons Attribution (CC BY) licence to any Author Accepted Manuscript version arising from this submission.

## REFERENCES


[1] Fontana, M.G. & Greene, N.D., *Corrosion Engineering*, McGraw-Hill Book Company, 1978.

[2] Bockris, J.O. & Reddy, A.K., *Modern Electrochemistry*, Springer Science and Business Media, 1974.

[3] Zamani, N., Porter, J. & Mufti, A., A survey of computational efforts in the field of corrosion engineering. *International Journal for Numerical Methods in Engineering*, **23**, pp. 1295–1311, 1986.

[4] Vogelius, M. & Xu, J.M., A nonlinear elliptic boundary value problem related to corrosion modeling. *Quarterly of Applied Mathematics*, **56**, pp. 479–505, 1998.

[5] Bryan, K. & Vogelius, M., Singular solutions to a nonlinear elliptic boundary value problem originating from corrosion modeling. *Quarterly of Applied Mathematics*, **60**, pp. 675–694, 2002.

[6] Jaswon, M.A. & Symm, G.T., *Integral Equation Methods in Potential Theory and Elastostatics*, Academic Press, 1977.

[7] Karageorghis, A. & Lesnic, D., The method of fundamental solutions for steady-state heat conduction in nonlinear materials. *Communications in Computational Physics*, **4**, pp. 911–928, 2008.

[8] Hansen, P.C., *Rank Deficient and Discrete Ill-Posed Problems: Numerical Aspects of Linear Inversion*, SIAM, 1998.

[9] Hansen, P.C., Analysis of discrete ill-posed problems by means of the L-curve. *SIAM Review*, **34**, pp. 561–580, 1992.

[10] Sincich, E., Stability and reconstruction for inverse corrosion problems. *Journal of Inverse and Ill-Posed Problems*, **17**, pp. 783–794, 2009.

[11] Rundell, W., Recovering an obstacle and its impedance from Cauchy data. *Inverse Problems*, **24**, pp. 783–794, 045003, 2008.

[12] Choulli, M., An inverse problem in corrosion detection: Stability estimates. *Journal of Inverse and Ill-Posed Problems*, **26**, pp. 453–457, 2018.

[13] Jin, B. & Jun, Z., Numerical estimation of the Robin coefficient in a stationary diffusion equation. *IMA Journal of Numerical Analysis*, **30**, pp. 677–701, 2010.

[14] Fasino, D. & Inglese, G., Recovering unknown terms in a nonlinear boundary condition for Laplace's equation. *IMA Journal of Applied Mathematics*, **71**, pp. 832–852, 2006.
[15] Fasino, D. & Inglese, G., Recovering nonlinear terms in an inverse boundary value problem for Laplace's equation: A stability estimate. *Journal of Computational and Applied Mathematics*, **198**, pp. 460–470, 2007.
[16] Cao, H., Pereverzev, S.V. & Sincich, E., Natural linearization for corrosion identification. *Journal of Physics: Conference Series*, **135**, 012027, 2008.
[17] Cakoni, F. & Kress, R., Integral equations for inverse problems in corrosion detection from partial Cauchy data. *Inverse Problems and Imaging*, **1**, pp. 229–245, 2007.
[18] Bacchelli, V., Uniqueness for the determination of unknown boundary and impedance with the homogeneous Robin condition. *Inverse Problems*, **25**, 015004, 2009.

# DISCONTINUOUS BOUNDARY CONDITION IMPACT ON NEAR CORNER STRESS VALUES AROUND A SPHERICAL CAVITY

AHMADREZA HABIBI
Norwegian University of Science and Technology, Norway

ABSTRACT
Numerical solutions for solid mechanics problems with a rapid variation of quantities in a small domain may result in impractical estimates. The presence of a cavity in an elastic domain, as a classic example, enforces more complexity on a large model due to its stress concentration impact. Boundary element method (BEM) as a numerical method applicable to this type of problem, also involves singular kernels. Due to the implementation complexity of augmented integration schemes, there is still a tendency to use ordinary integration techniques. Besides, enforcing boundary conditions might also be challenging in BEM. In this study, near-corner stress values are explored to evaluate the impact of straightforward integration schemes and discontinuous boundary enforcement treatments. Provided results demonstrate that for coordinates near corners in this specific problem, using non-conforming elements and shape functions with displacement vectors might reduce the error values to fewer than 1%. Further, despite the results for nodes away from corners, unique traction can compromise accuracy for near-corner coordinates in exchange for its simple implementation.
*Keywords: boundary element method, non-conforming elements, corner points.*

## 1 INTRODUCTION

Incorporating analytical solutions, boundary element method (BEM) has the capacity to provide precise results for a problem with high field variable gradients. Considerably, BEM weaknesses regarding computer memory and processing speed have been improved using computing algorithms, larger built-in memory, and parallel codes; however, the presence of challenges like singular integrands and corner points still might be problematic [1]–[6].

The traction vector in applied BEM formulation has been derived using the analytical solution for displacement and elasticity theory; therefore, it is dependent on the normal vectors of the boundary surface, and it causes difficulties in implementation for discontinuous boundary conditions and corner points. Among many suggested methods to address the issue, rounding off the boundary, extra degrees of freedom [1], [7], [8], unique traction [9], [10], additional collocation points [5], [11], [12], non-conforming elements (applied here) [2], [13]–[18] can be mentioned.

In the present work, elastostatic BEM code results have been evaluated for a problem with stress concentration. A classic problem with a high gradient of field variables and an available analytical solution is the spherical cavity inside an elastic solid [19], [20]. In previous research outputs [9], [21], using the same problem, the accumulated stress error values away from the corners have been presented. In this study, the error values near corners have been calculated separately for each internal position for different non-conforming elements, mesh, and integration points by employing simple integration and approximation methods.

## 2 GOVERNING EQUATIONS AND METHODS

The basic BEM formulation is mentioned in eqn (1).

$$C_{ij}u_j(P) = \int_\Gamma U_{ij}(Q,P)\, t_j(Q)\, d\Gamma(Q) - \oint_\Gamma T_{ij}(Q,P)\, u_j(Q)\, d\Gamma(Q). \quad (1)$$

WIT Transactions on Engineering Sciences, Vol 136, © 2024 WIT Press
www.witpress.com, ISSN 1743-3533 (on-line)
doi:10.2495/BE470031

Terms $U$ and $T$ are well-known singular functions from Kelvin's solution, $\Gamma$ is the boundary surface, and the second integral on the right sight of the equations implies that the Cauchy principal value is applied. Moreover, $\delta_{ij}$, $u$, and $t$ are the Kronecker delta function, displacement, and traction variables, respectively. Also, $P$ and $Q$ represent source and field points [1], [3]. As free term ($C_{ij}$) can not be calculated easily using direct methods, widely utilised free rigid body motion has been used [22]; this also provides accurate diagonal arrays of the matrix $T$.

The only augmented integration scheme used in this study is the degenerate mapping of a triangle to a standard quadrilateral element which has a zero Jacobian of mapping [1], [23]. This approach was applied for coordinates close to an element surface (critical distance was set to a few times more than the largest element length [24]). For coordinates close to an element, division to triangles and degenerate mapping were done using the closet position on the element. Moreover, results of this study have been derived by 6 and 12 Gaussian quadrature in each direction.

To tackle the discontinuous boundary conditions, unique tractions (UT), semi-discontinuous element (SDE), improved semi-discontinuous element (SDEI), and discontinuous element (DE) have been used separately. While UT considers a unique traction vector as the variable and enforces extra specified traction boundary conditions using a separate constant vector, other methods generate new nodes at specific locations and introduce new degrees of freedom to differentiate between variables where it is mathematically required. Elaborating, DE displaces all nodes even if one of them has discontinuous traction values. The only difference between SDE and SDEI is that in SDEI, if it is possible, nodes are moved along the edges rather than into the element surface. Therefore, for all methods, the integration surface is similar, while nodal positions are different [21].

Stress error calculation was performed using an analytical solution provided and utilised for a cavity inside an infinite elastic domain [19]–[21]. By choosing a proper model size, the solution result can be applied for finite domain. Further, error value was derived using relative error equation for each coordinate separately. As the simple integration scheme applied here cannot tackle the strong singularity for nodes near the boundary, in the post-processing section, nodal positions of hexahedral elements for stress and subsequently error calculation were considered; this allowed the program to use shape functions and the BEM stress equation as two different ways of stress calculation.

## 3 NUMERICAL RESULTS

Using a similar model to previous works [9], [21], a cavity inside a cubic elastic body was considered. Benefiting from the symmetrical feature of the problem, 1/8 of the model was designed to be meshed and imported as the imported data file. The cube side length was big enough to avoid the boundary effect around the cavity. A tensile surface force of 1 MN on top (positive Z-direction) and proper boundary conditions on other surfaces were enforced (see Fig. 1). Two sets of hexahedral elements, so close to corners, were considered for post-processing; the stress values for their nodes were calculated with both the BEM solution and shape functions and compared with the analytical solution to produce error values for each coordinate separately (see Fig. 1 and Table 1).

## 4 RESULTS AND DISCUSSION

Stress error value results have been depicted in Fig. 3. Due to the high error values at some cases and coordinates, all error values more than an acceptable value (1%) are shown by one

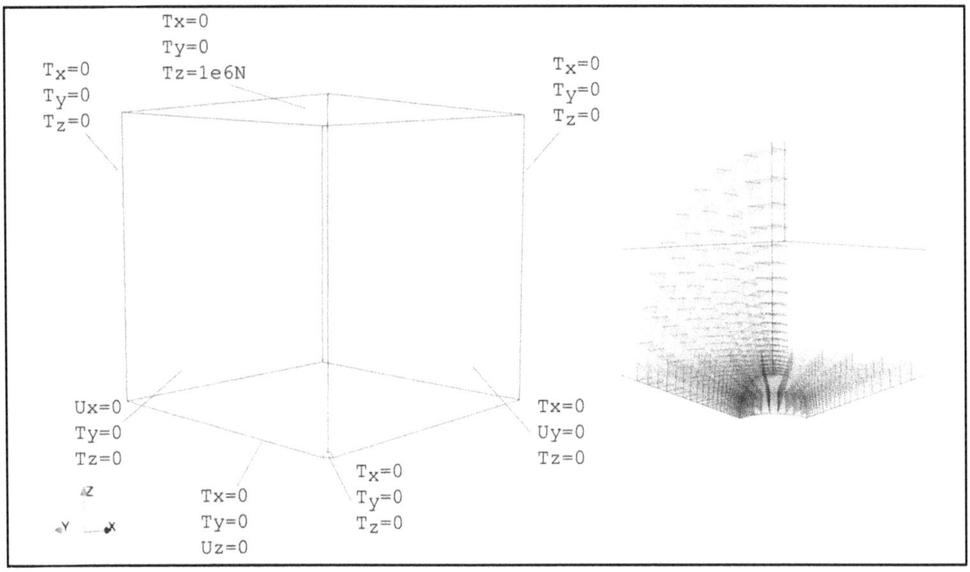


Figure 1: An outline of the model and boundary conditions (left) and post-processing elements' nodal (right).

Table 1: Analysis input data.

| **Material properties** | Young's modulus (Pa) | Poisson's ratio | | |
|---|---|---|---|---|
| | 1.00E+09 | 0.3 | | |
| **BEM mesh** | Element length (m) | | | |
| | Cavity surface | Lateral surface | | |
| 1 | ~0.18 | 0.2<L<0.28 | | |
| 2 | ~0.09 | 0.1<L<0.18 | | |
| **Post-processing sets** | Set position and distance to the boundary | | Distance ratio to the BEM element length | |
| | Parallel to boundary plane/ off-set (m) | Distance to the cavity surface (m) | Cavity elements | Parallel boundary surface elements |
| 1 | X–Y plane/0.02 | 0.02 | ~0.11 | ~0.1 |
| 2 | Y–Z plane/0.02 | 0.02 | ~0.22 | ~0.2 |

colour, which is a sign of computation method failure (applied to all figures). Therefore, this comparison is focused on the error values lower than 1%. Also, DE and SDEI results were not included in the figure due to the similarity to SDE results. Furthermore, Fig. 2 illustrates different nodal coordinates in various methods. Worth mentioning that despite previous work [9], this study is more focused on separated error values for coordinates near corners.

Due to the simple integration scheme, the BEM stress equation deviated from the solution at the nodes close to the boundary. However, noticeably, high error values were mostly at the

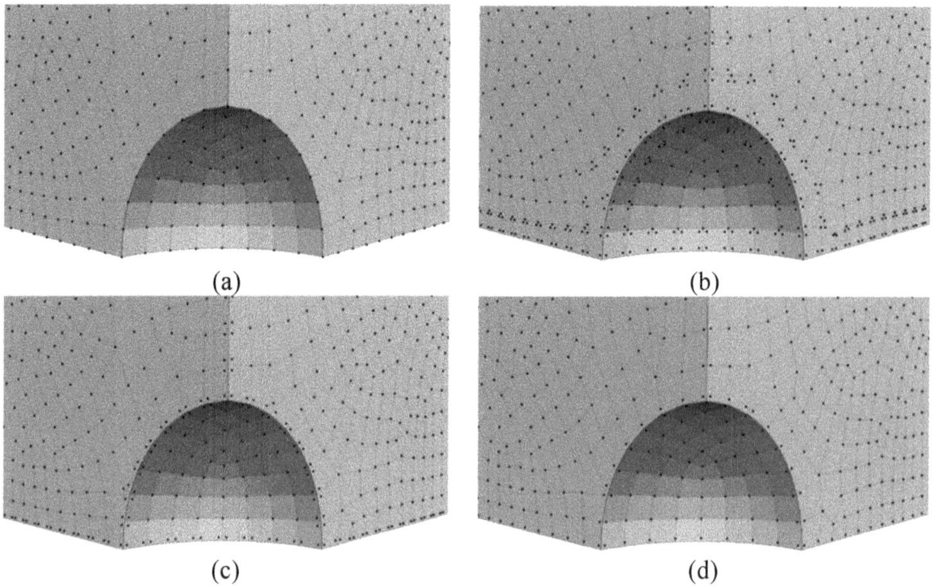

Figure 2: Different nodal positions in (a) UT, (b) DE, (c) SDE, and (d) SDEI.

first layer of nodes, while the distance between the two closest layers to the boundary is 0.05 m, which implies that boundary layer impact has disappeared at a few fractions of element length. Moreover, refining the mesh to half the size has provided minor improvements for the BEM equation and slightly better results for nodes near the cavity at the bottom face (see for example Fig. 3(b) and (f)).

Interestingly, using shape functions in UT for both meshes provided inaccurate stress values near corners and more precise values for nodes along the boundaries away from corners (see Fig. 3(a) and (e)). When non-conforming elements were utilised, hexahedral shape functions provided more accurate results at nearly all nodal coordinates for both meshes (see Fig. 3(c)). However, none of the approaches eliminated the errors for the BEM stress formulation raised due to integration (see Fig. 3(b), (d), and (f)).

All non-conforming elements' results were regenerated with two higher deviation ratio constants (ratio of the distance at which nodes were displaced to make non-conforming elements) for both meshes. The comparison of accumulated error values suggested that the change in error values was quite negligible. However, for stress values calculated using shape functions, compared to the other two methods, there was a slightly more consistent increase trend in SDEI error values by increasing the degree of discontinuity (see Fig. 4). Additionally, a more quantitative comparison showed that although the accumulated error values of the non-conforming methods were so close, the highest point-wise error values limit for SDE and SDEI, calculated using shape functions, were slightly lower than DE.

Furthermore, simulations were reprocessed by increasing the number of Gaussian quadrature points from six to 12 in each direction. The results indicated slight improvement in error values calculated using shape functions. However, for the BEM formulation, a noticeable reduction in error values was observed for coordinates near the boundaries. Interestingly, the degree of improvement was less pronounced for the coarse mesh, likely due

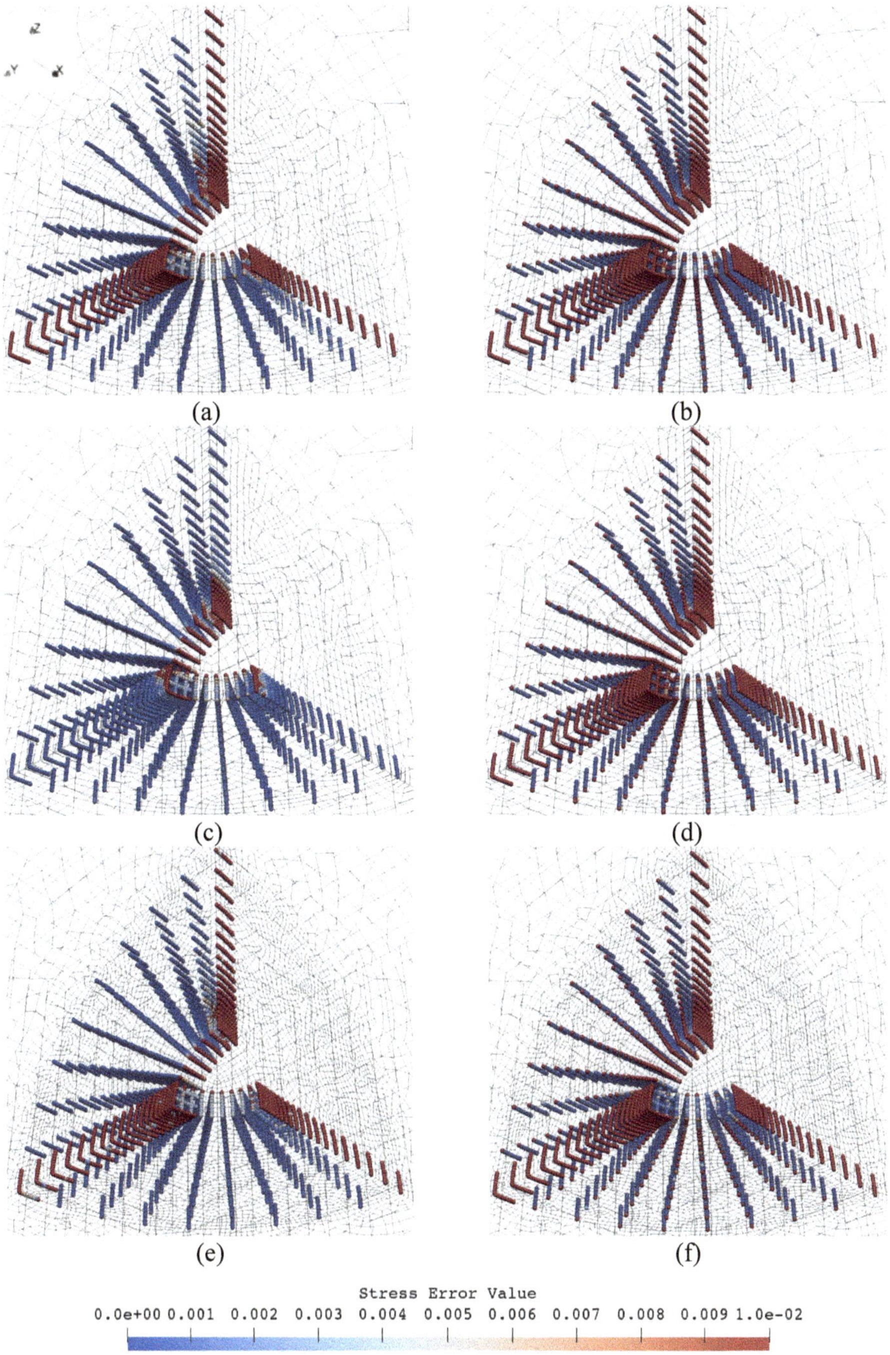


Figure 3: Nodal error values of stress tensor calculated in post-processing using element shape functions (left panel) and the BEM stress formula (right panel) for both coarse and fine meshes. (a) UT-coarse mesh; (b) UT-coarse mesh; (c) SDE-coarse mesh; (d) SDE-coarse mesh; (e) UT-fine mesh; and (f) UT-fine mesh.

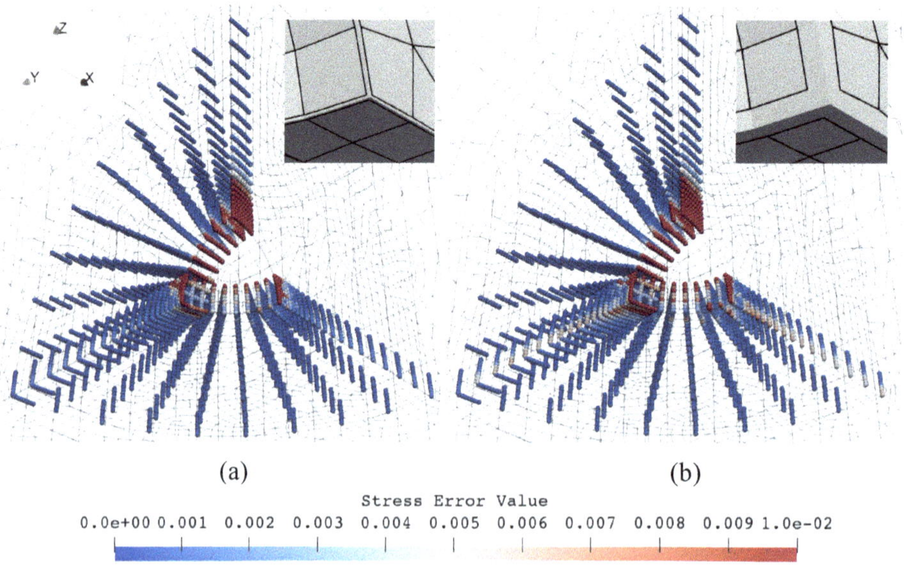


Figure 4: Stress error values calculated using shape functions and SDEI elements with higher deviation ratio in (b).

to the stronger boundary layer effect associated with bigger elements (see Fig. 5). To further reduce the boundary layer effects, the model was meshed with 0.05 m elements. Despite this refinement, still for some coordinates close to the boundary (0.02 m distance) high BEM formula error values were detected (Fig. 6). Thus, increasing the number of collocation and integration points without adequately addressing higher-order BEM stress formula singularities in this range of distance ratios may not eliminate the risk of erroneous results.

## 5 CONCLUDING REMARKS

In this study, the impact of the boundary layer and non-conforming elements on the accuracy of discontinuous traction vectors and stress function singularity was examined.

Based on the results provided using an elastostatic BEM code, it was shown that for this specific problem, non-conforming elements can yield accurate results when post-processing coordinates are placed near corners. Furthermore, even with simple integration schemes, the BEM stress equation can still deliver accurate results for coordinates just a few element-length fractions away from the boundary. These findings align with the previous research outputs on coordinates away from corners.

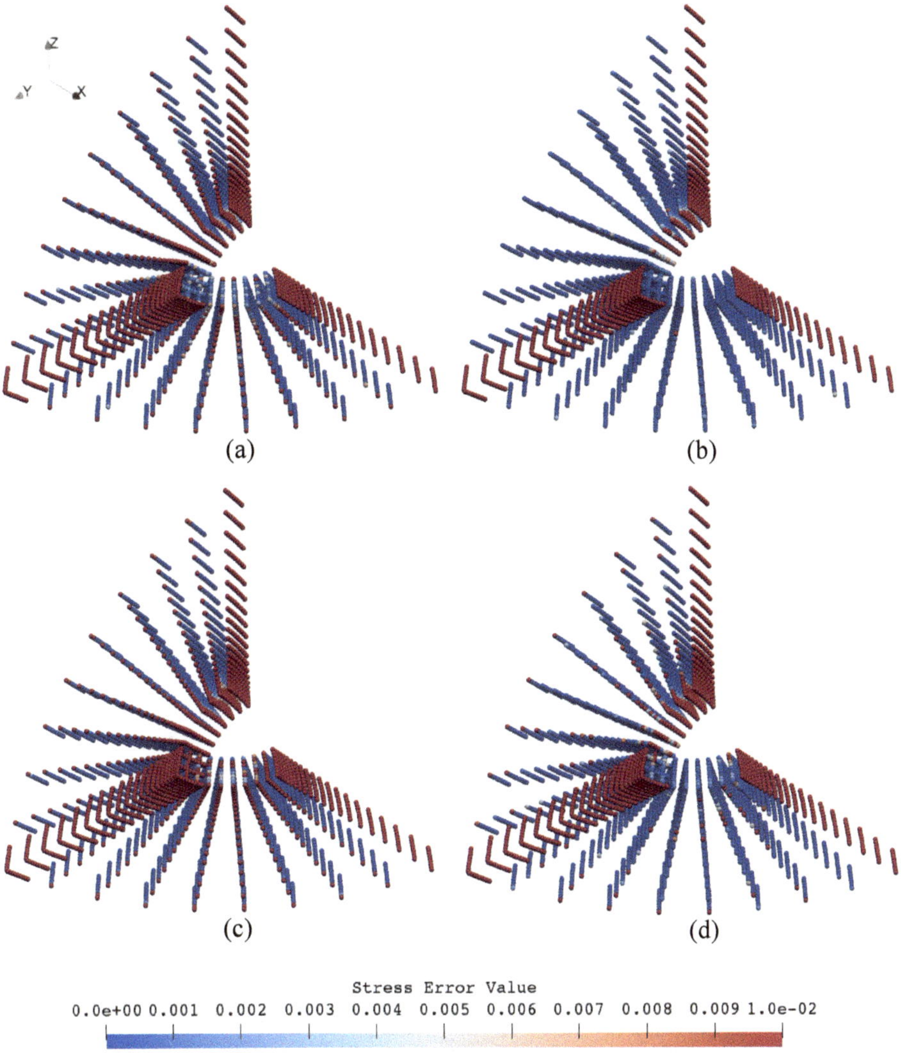


Figure 5: Change in BEM formula stress error values with an increase in the number of Gaussian quadrature points from six (left panel) to 12 (right panel). (a) and (b) show results for the fine mesh; (c) and (d) show the results for the coarse mesh.

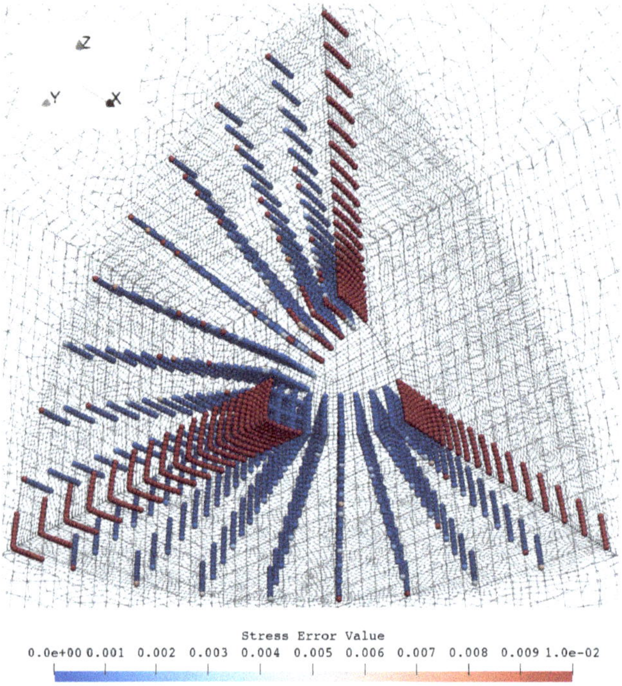


Figure 6: BEM stress formulation error values for SDE method and an element size of 0.05 m in the vicinity of the cavity.

## REFERENCES

[1] Gao, X.-W. & Davies, T.G., *Boundary Element Programming in Mechanics*, Cambridge University Press, 2002.

[2] Aliabadi, M.H., *The Boundary Element Method, Volume 2: Applications in Solids and Structures*, Wiley, 2002.

[3] Crouch, S.L., Starfield, A.M. & Rizzo, F., *Boundary Element Methods in Solid Mechanics*, 1983.

[4] Alarcon, E., Brebbia, C. & Dominguez, J., The boundary element method in elasticity. *International Journal of Mechanical Sciences*, **20**(9), pp. 625–639, 1978.

[5] Leng, D., *Boundary Element Method in Anisotropic Media with Grain Sliding and Dislocation Dynamics*, Iowa State University, 2007.

[6] Hall, W.S., Boundary element method. *The Boundary Element Method*, Springer, pp. 61–83, 1994.

[7] Gao, X.-W. & Davies, T., 3D multi-region BEM with corners and edges. *International journal of Solids and Structures*, **37**(11), pp. 1549–1560, 2000.

[8] Qin, X., Lei, W., Li, H. & Fan, Y., Corner treatment in 3D time-domain boundary element method. *Journal of the Brazilian Society of Mechanical Sciences and Engineering*, **44**(11), p. 556, 2022.

[9] Habibi, A., Numerical stress computation around cavity using boundary element method for elastic material. *58th US Rock Mechanics/Geomechanics Symposium*, 2024.

[10] Cruse, T.A., An improved boundary-integral equation method for three dimensional elastic stress analysis. *Computers and Structures*, **4**(4), pp. 741–754, 1974.
[11] Mitra, A.K. & Ingber, M.S., A multiple-node method to resolve the difficulties in the boundary integral equation method caused by corners and discontinuous boundary conditions. *International Journal for Numerical Methods in Engineering*, **36**(10), pp. 1735–1746, 1993.
[12] Wang, Z. & Wu, Q., A new approach treating corners in boundary element method. *Boundary Elements XIII*, Springer, pp. 901–911, 1991.
[13] Dyka, C. & Millwater, H., Formulation and integration of continuous and discontinuous quadratic boundary elements for two dimensional potential and elastostatics. *Computers and Structures*, **31**(4), pp. 495–504, 1989.
[14] Yan, G. & Lin, F.-B., Treatment of corner node problems and its singularity. *Engineering Analysis with Boundary Elements*, **13**(1), pp. 75–81, 1994.
[15] Mi, Y. & Aliabadi, M., Dual boundary element method for three-dimensional fracture mechanics analysis. *Engineering Analysis with Boundary Elements*, **10**(2), pp. 161–171, 1992.
[16] Parreira, P., On the accuracy of continuous and discontinuous boundary elements. *Engineering Analysis*, **5**(4), pp. 205–211, 1988.
[17] Okada, H., Rajiyah, H. & Atluri, S.N., A novel displacement gradient boundary element method for elastic stress analysis with high accuracy, 1988.
[18] Manolis, G. & Banerjee, P., Conforming versus non-conforming boundary elements in three-dimensional elastostatics. *International Journal for Numerical Methods in Engineering*, **23**(10), pp. 1885–1904, 1986.
[19] Neuber, H., *Theory of Notch Stresses: Principles for Exact Stress Calculation*, JW Edwards, 1946.
[20] Brethauer, G. (ed.), Stress around pressurized spherical cavities in triaxial stress fields. *International Journal of Rock Mechanics and Mining Sciences and Geomechanics Abstracts*, Elsevier, 1974.
[21] Habibi, A., Comparison of boundary element numerical techniques for corner point treatment. Submitted for publication.
[22] Brebbia, C.A., Telles, J.C.F. & Wrobel, L.C., *Boundary Element Techniques: Theory and Applications in Engineering*, Springer Science and Business Media, 2012.
[23] Lean, M.H. & Wexler, A., Accurate numerical integration of singular boundary element kernels over boundaries with curvature. *International Journal for Numerical Methods in Engineering*, **21**(2), pp. 211–228, 1985.
[24] Telles, J., A self-adaptive co-ordinate transformation for efficient numerical evaluation of general boundary element integrals. *International Journal for Numerical Methods in Engineering*, **24**(5), pp. 959–973, 1987.

# SECTION 2 COMPUTATIONAL METHOD AND OTHERS

# HIGH-ACCURATE APPROACH FOR REPEATED INTEGRALS CALCULATION FROM DOUBLE LAYER POTENTIAL WITH PIECEWISE-CONSTANT DENSITY OVER TRIANGULAR PANELS

ILIA K. MARCHEVSKY[1] & SOPHIA R. SERAFIMOVA[2]
[1]Bauman Moscow State Technical University, Russia
[2]Kazan Federal State University, Russia

## ABSTRACT

The algorithm of vortex particle methods of computational hydrodynamics for three-dimensional incompressible flows simulation includes the solution of a boundary integral equation (BIE) that describes vorticity generation on the body surface. Instead of the 'traditional' approach, which leads to hyper-singular integral equation, an alternative way allows for considering the BIE with weak singularity, that can be solved with acceptable accuracy by using the Galerkin approach with piecewise-constant solution representation. The kernel of the BIE is the gradient of fundamental solution of the Laplace equation. For the panels with common edge or common vertices high-accurate algorithm is developed, that is based on singularity additive exclusion and its analytical integration. The resulting formulae are written down in such a way to avoid accumulation of roundoff errors. The special cases are considered for which ambiguities arise: limit values of the corresponding terms are calculated analytically and linear expansions are presented in neighbourhood of 'ambiguous cases' with respect to small parameters which use allows for achieving high accuracy of repeated integrals computation.
*Keywords: vortex particle method, boundary integral equation, Galerkin approach, repeated integral, singularity exclusion, common edge, common vertex, limit value, linear expansion.*

## 1 INTRODUCTION

The boundary element method [1]–[3] is a powerful tool for solving a wide class of engineering problems. The main object here is boundary integral equations (BIEs), which solution makes it possible to find solutions of boundary value problems for partial differential equations of elliptic type. In this paper, the problem is considered that arises when implementing vortex particle methods [4], [5] of computational fluid dynamics when modelling the spatial flow of an incompressible medium around bodies.

Discrete analogues of BIEs are systems of linear algebraic equations with dense matrices. Despite the significant amount of research carried out by various authors concerning the construction of discrete analogues of BIEs, this issue continues to attract attention; in particular, it is of interest to construct numerical schemes based on the Galerkin approach. As numerical experiments show, schemes of such type turn out to be very accurate even with a piecewise constant representation of the solution and the use of low-quality surface meshes [6]. The main difficulty in this case is the need to calculate repeated integrals of the BIE kernel function over parts (panels) of the surface: integrals of unbounded functions that can be improper, singular or hyper-singular in cases where the panels have common points. To calculate such integrals, one can use the Taylor–Duffy approach [7], which is quite general, but the corresponding computational subroutine turns out to be quite time-consuming. The same applies to the methodology proposed in Dodig et al. [8].

The aim of this paper is to develop an algorithm for calculating with high accuracy the integrals that arise at construction of a discrete analogue of the BIE with a piecewise constant representation of the solution and a triangulated representation of the domain boundary. The obtained results can be applicable for the problems, where there is a need of solving the Laplace equation, and solution of the Dirichlet/Neumann problem is represented through a double/single layer potential, respectively. Similar integrals arise when a vortex sheet is considered on the body surface and its influence is calculated [9].

## 2 INTEGRAL CALCULATION OVER TRIANGULAR PANEL

Under assumption of piecewise-constant representation of the solution (single, double or vortex layer density) over triangulated body surface, the problem of discretisation of the governing boundary integral equation using the simplest numerical scheme – the collocation method – can be formulated as a problem of integration of the gradient of fundamental solution of the Laplace equation over triangular panel.

Such integral can be calculated exactly for arbitrary triangle $K_j$ and arbitrary observation point $M_i$ which position is $\mathbf{r}_i$:

$$\mathbf{J}(M_i, K_j) = \int_{K_j} \frac{\mathbf{r}_i - \boldsymbol{\xi}}{4\pi |\mathbf{r}_i - \boldsymbol{\xi}|^3} dS_\xi = \frac{1}{4\pi}\left(\Theta(\mathbf{v}_a, \mathbf{v}_b, \mathbf{v}_c)\mathbf{n}_j + \boldsymbol{\Psi}(\mathbf{v}_a, \mathbf{v}_b, \mathbf{v}_c) \times \mathbf{n}_j\right). \qquad (1)$$

Here

$$\Theta(\mathbf{v}_a, \mathbf{v}_b, \mathbf{v}_c) = 2\arctan\left(\mathbf{v}'_a \mathbf{v}'_b \mathbf{v}'_c,\ 1 + (\mathbf{v}'_a \cdot \mathbf{v}'_b) + (\mathbf{v}'_b \cdot \mathbf{v}'_c) + (\mathbf{v}'_c \cdot \mathbf{v}'_a)\right),$$

$$\boldsymbol{\Psi}(\mathbf{v}_a, \mathbf{v}_b, \mathbf{v}_c) = \ln\left(\frac{|\mathbf{v}_a|}{|\mathbf{v}_b|}\frac{1+\cos\varphi_c^a}{1-\cos\varphi_c^b}\right)\boldsymbol{\tau}_c + \ln\left(\frac{|\mathbf{v}_b|}{|\mathbf{v}_c|}\frac{1+\cos\varphi_a^b}{1-\cos\varphi_a^c}\right)\boldsymbol{\tau}_a + \ln\left(\frac{|\mathbf{v}_c|}{|\mathbf{v}_a|}\frac{1+\cos\varphi_b^c}{1-\cos\varphi_b^a}\right)\boldsymbol{\tau}_b,$$

the function $\arctan(y, x)$ has values in range $(-\pi, \pi]$ and means the principal value of the argument of complex number $z = x + iy$.

The quantities included in eqn (1) have a clear geometric meaning (Fig. 1): the primes indicate the unit vectors of the corresponding vectors; $\mathbf{v}_0 = 1/3(\mathbf{v}_a + \mathbf{v}_b + \mathbf{v}_c)$; vector $\mathbf{n}_j$ is the unit vector of the normal to the plane of the triangle, that should be oriented in such a way that when observing the triangle against the direction of this vector, the sides having lengths $L_a$, $L_b$ and $L_c$ are traversed counterclockwise; $\Theta(\mathbf{v}_a, \mathbf{v}_b, \mathbf{v}_c)$ is a value equal in modulus to the solid angle at which the triangle is visible from the observation point, the expression for calculating the module of the solid angle is borrowed from van Oosterom and Strackee [10], while $\Theta(\mathbf{v}_a, \mathbf{v}_b, \mathbf{v}_c) > 0$ if the observation point lies from the panel in the direction of the normal vector $\mathbf{n}_j$, i.e., at $\mathbf{v}_0 \cdot \mathbf{n}_j > 0$; $\boldsymbol{\tau}_a$, $\boldsymbol{\tau}_b$ and $\boldsymbol{\tau}_c$ are unit vectors that correspond to the vectors directed along the sides of the triangle $K_j$.

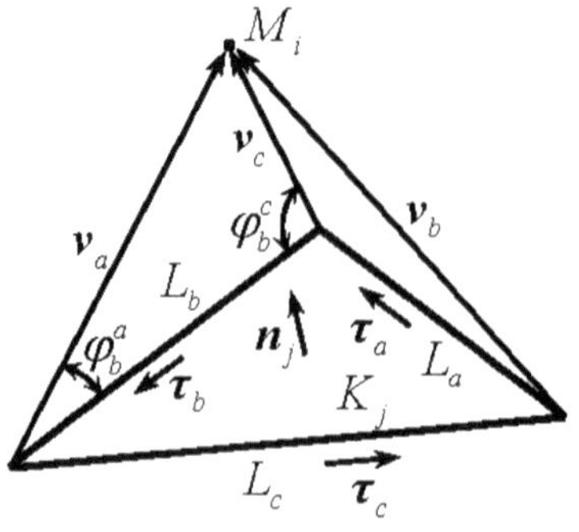


Figure 1: Observation point $M_i$, triangular panel $K_j$ and all necessary angles and vectors.

Let us consider special cases when eqn (1) cannot be applied straightforwardly.

1. If the observation point $M_i$ lies inside the influencing panel $K_j$, which in practical calculations, depending on the situation, can either be postulated explicitly or require checking the condition that the distance from the point to the panel is small compared to the linear size of the panel, then the expression (1) remains valid if one takes $\Theta(\mathbf{v}_a, \mathbf{v}_b, \mathbf{v}_c) = 0$. The limit values of the integral on both sides of the panel $\mathbf{J}^{\lim}(M_i, K_j)$ are obtained by taking $\Theta(\mathbf{v}_a, \mathbf{v}_b, \mathbf{v}_c) = \pm 2\pi$; the sign is determined by the side from which the observation point approaches the panel.
2. When the observation point $M_i$ lies in the plane of the triangular panel on the continuation of one of the sides of the triangle in the direction opposite to the corresponding vector $\boldsymbol{\tau}$, i.e., when one of the following conditions is satisfied,

$$\cos\varphi_c^a = -1, \quad \cos\varphi_a^b = -1, \quad \cos\varphi_b^c = -1,$$

an ambiguity arises in the corresponding term in the expression for $\boldsymbol{\Psi}(\mathbf{v}_a, \mathbf{v}_b, \mathbf{v}_c)$. In such situations, the logarithm that contains ambiguity should be replaced with its limit value:

$$\lim_{\cos\varphi_c^a \to -1} \ln\left(\frac{|\mathbf{v}_a|}{|\mathbf{v}_b|}\frac{1+\cos\varphi_c^a}{1-\cos\varphi_c^b}\right) = -\ln\frac{|\mathbf{v}_a|}{|\mathbf{v}_b|}, \qquad \lim_{\cos\varphi_a^b \to -1} \ln\left(\frac{|\mathbf{v}_b|}{|\mathbf{v}_c|}\frac{1+\cos\varphi_a^b}{1-\cos\varphi_a^c}\right) = -\ln\frac{|\mathbf{v}_b|}{|\mathbf{v}_c|},$$

$$\lim_{\cos\varphi_b^c \to -1} \ln\left(\frac{|\mathbf{v}_c|}{|\mathbf{v}_a|}\frac{1+\cos\varphi_b^c}{1-\cos\varphi_b^a}\right) = -\ln\frac{|\mathbf{v}_c|}{|\mathbf{v}_a|}.$$

Note that practical calculations are inevitably accompanied by errors that are caused, at least, by the finite precision of the representation of real numbers, so equalities of the type $\cos\varphi_c^a = -1$ will not be satisfied exactly in the mathematical sense (with very rare exceptions). At the same time, operations with a value of $\cos\varphi_c^a$ close to $(-1)$ will lead to a rapid accumulation of errors due to the subtraction of close numbers. Thus, in practice it is necessary to enter a small value of 'angular tolerance' $\varepsilon > 0$ and check the conditions

$$1 + \cos\varphi_c^a < \frac{\varepsilon^2}{2}, \qquad 1 + \cos\varphi_a^b < \frac{\varepsilon^2}{2}, \qquad 1 + \cos\varphi_b^c < \frac{\varepsilon^2}{2},$$

meaning that the angles $\varphi_c^a$, $\varphi_a^b$ and $\varphi_b^c$ differ from the flat angle by less than $\varepsilon$. For small values of $\varepsilon$ the limit values differ from the exact ones by $O(\varepsilon^2)$.

The only case when the integral $\mathbf{J}(M_i, K_j)$ does not converge arises if the observation point $M_i$ lies on sides of the panel $K_j$.

## 3 REPEATED INTEGRAL CALCULATION OVER TWO TRIANGULAR PANELS

First of all, let us note, that it is impossible to obtain exact analytical expressions in closed form for repeated integrals form the gradient of Newtonian potential over two triangular panels. Numerical integration over the control panel $K_i$ using Gaussian (or some other) quadrature rules [11] can be done only for the cases when the panels $K_i$ and $K_j$ do not have common points: the results of calculating the internal integral eqn (1) is now a bounded function together with their derivatives with respect to spatial coordinates, therefore, high accuracy of calculating the external integral normally can be achieved when using 'standart' quadrature rules, even with a relatively small number of nodes.

If the panels $K_i$ and $K_j$ have a common edge or vertex, the internal integral, i.e., the function $\mathbf{J}(M_i, K_j)$, turns out to be unbounded when approaching the common points, and the external integral becomes improper. To calculate it with high accuracy, it is suggested to perform singularity extraction in the integrand, i.e., to write it down as a sum

$$\mathbf{J}(M_i, K_j) = \mathbf{J}^{\text{reg}}(M_i, K_j) + \mathbf{J}^{\text{sing}}(M_i, K_j),$$

where the singular part includes the singular parts of both functions $\Theta$ and $\mathbf{\Psi}$:

$$\mathbf{J}^{\text{sing}}(M_i, K_j) = \frac{1}{4\pi}\left(\Theta^{\text{sing}}(M_i, K_j)\mathbf{n}_j + \mathbf{\Psi}^{\text{sing}}(M_i, K_j) \times \mathbf{n}_j\right).$$

Note, that $\Theta(M_i, K_j)$ is always bounded, but has unbounded first derivations which, in fact, are 'isolated' in singular part $\Theta^{\text{sing}}(M_i, K_j)$. Then the regular part (dependencies on the position of observation point $M_i$ and panel $K_j$ are omitted)

$$\mathbf{J}^{\text{reg}} = \frac{1}{4\pi}\left((\Theta - \Theta^{\text{sing}})\mathbf{n}_j + (\mathbf{\Psi} - \mathbf{\Psi}^{\text{sing}}) \times \mathbf{n}_j\right),$$

and also its spatial derivatives turn out to be smooth bounded functions, so their integration can be performed numerically, if no one of the nodes of the used quadrature rule coincide with the common points of the considered panels. It introduces a non-burdensome restriction for the quadrature rule.

For integrals from singular parts of the corresponding functions

$$\int_{K_i} \Theta^{\text{sing}} dS_r, \qquad \int_{K_i} \mathbf{\Psi}^{\text{sing}} dS_r$$

it is possible to derive exact expressions in closed form, which differ significantly for the cases of panels with a common edge and a common vertex.

Note that if the control and influencing panels coincide ($i = j$), then the repeated integral becomes singular; its principal value in the Cauchy sense is equal to zero.

## 3.1 Panels with common edge

Let us consider the control panel $K_i$ and the influencing panel $K_j$ with common edge, which directing unit vector is $\boldsymbol{\tau}_c$ (Fig. 2).

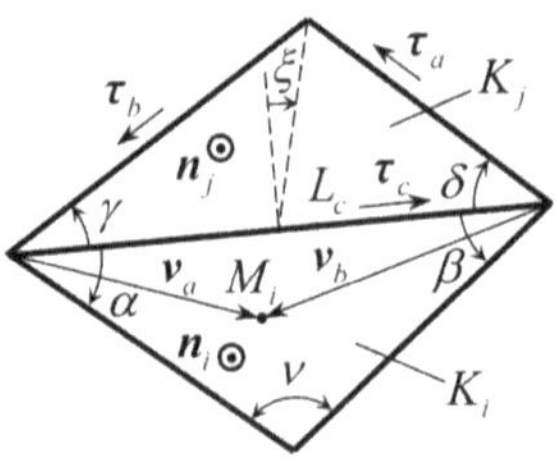


Figure 2: Two panels with common edge.

The expressions for singular parts of the inner integral $\mathbf{J}(M_i, K_j)$ are the following:

$$\Theta^{\text{sing}}(M_i, K_j) = 2\left(\arctan\left(\mathbf{v}'_a\boldsymbol{\tau}_b\boldsymbol{\tau}_c,\ (\boldsymbol{\tau}_b - \boldsymbol{\tau}_c)\cdot(\boldsymbol{\tau}_b + \mathbf{v}'_a)\right) - \arctan\left(\mathbf{v}'_b\boldsymbol{\tau}_a\boldsymbol{\tau}_c,\ (\boldsymbol{\tau}_a - \boldsymbol{\tau}_c)\cdot(\boldsymbol{\tau}_a - \mathbf{v}'_b)\right)\right),$$

$$\mathbf{\Psi}^{\text{sing}}(M_i, K_j) = \boldsymbol{\tau}_c \ln\left(\frac{|\mathbf{v}_b|\,\boldsymbol{\tau}_c\cdot(\boldsymbol{\tau}_c - \mathbf{v}'_b)}{|\mathbf{v}_a|\,\boldsymbol{\tau}_c\cdot(\boldsymbol{\tau}_c - \mathbf{v}'_a)}\right) - \boldsymbol{\tau}_b \ln\left(\frac{|\mathbf{v}_a|}{L_c}\boldsymbol{\tau}_b\cdot(\boldsymbol{\tau}_b + \mathbf{v}'_a)\right)$$
$$-\boldsymbol{\tau}_a \ln\left(\frac{|v_b|}{L_c}\boldsymbol{\tau}_a\cdot(\boldsymbol{\tau}_a - \mathbf{v}'_b)\right).$$

The expression for $\Theta^{\text{sing}}$ as well as scalar multipliers for unit vectors $\boldsymbol{\tau}_p$ in expression $\boldsymbol{\Psi}^{\text{sing}}$ can be integrated analytically in closed form over the control panel $K_i$:

$$\int_{K_i} \Theta^{\text{sing}}(M_i, K_j) dS_r = S_i(q_\Theta(\xi, \alpha, \beta, \gamma, \mu, \lambda) + q_\Theta(\xi, \beta, \alpha, \delta, \sigma, \theta)), \tag{2}$$

$$\int_{K_i} \boldsymbol{\Psi}^{\text{sing}}(M_i, K_j) dS_r = S_i(q_\Psi(\xi, \alpha, \beta, \mu, \gamma, \lambda)\boldsymbol{\tau}_b$$
$$+ q_\Psi(\xi, \beta, \alpha, \sigma, \delta, \theta)\boldsymbol{\tau}_a - q_{\alpha\beta}(\alpha, \beta)\boldsymbol{\tau}_c). \tag{3}$$

Hereinafter $\alpha$ and $\beta$ mean the angles of the control panel $K_i$ adjuscent to the common edge; $\nu = \pi - \alpha - \beta$ is the opposite angle; $\gamma$ and $\delta$ mean the angles of the influencing panel $K_j$, that are adjuscent to the angles $\alpha$ and $\beta$, respectively; $S_i$ is the area of the panel $K_i$; $\xi$ is the angle between the planes of the panels $K_i$ and $K_j$, that can have both positive and negative values (it is measured as shown in Fig. 2):

$$\xi = \arctan(\mathbf{n}_i \mathbf{n}_j \boldsymbol{\tau}_c, \mathbf{n}_i \cdot \mathbf{n}_j),$$

the rest angles are defined as following:

$$\sigma = \pi - \arccos(\cos\alpha\cos\delta + \cos\xi\sin\alpha\sin\delta),$$
$$\mu = \pi - \arccos(\cos\beta\cos\gamma + \cos\xi\sin\beta\sin\gamma),$$
$$\lambda = \pi - \arccos(\cos\alpha\cos\gamma - \cos\xi\sin\alpha\sin\gamma),$$
$$\theta = \pi - \arccos(\cos\beta\cos\delta - \cos\xi\sin\beta\sin\delta),$$

which can be interpreted as a formulation of the cosine theorem for trihedral angles with the corresponding plane and dihedral angles.

The auxiliary functions $q_\Theta$, $q_\Psi$ and $q_{\alpha\beta}$, that depend only on the above mentioned angles, have the following expressions:

$$q_\Theta(\xi, \alpha, \beta, \gamma, \mu, \lambda) = \phi(\xi, \alpha, \gamma, \lambda) +$$
$$+ \frac{\sin\gamma\sin\nu}{\sin\alpha(1-\cos^2\mu)}\Big((\cos\beta\sin\gamma - \cos\xi\sin\beta\cos\gamma)\phi(\xi, \pi-\alpha, \pi-\gamma, \lambda) +$$
$$+ \frac{1}{2}\sin\xi\sin\beta\left((1+\cos\mu)\ln\frac{1+\cos\beta}{1-\cos\nu} + (1-\cos\mu)\ln\frac{1-\cos\beta}{1+\cos\nu} + 2\ln\frac{1+\cos\lambda}{1-\cos\gamma}\right)\Big), \tag{4}$$

$$q_\Psi(\xi, \alpha, \beta, \gamma, \mu, \lambda) = \frac{3}{2} - \frac{1}{\sin\alpha(1-\cos^2\mu)}\Big(\sin\beta(\cos\nu + \cos\mu\cos\lambda)\ln(1+\cos\lambda) +$$
$$+ \sin\nu(\cos\beta + \cos\mu\cos\gamma)\ln(1-\cos\gamma) + \sin\beta(1-\cos\mu)(\cos\nu - \cos\lambda)\ln\frac{\sin\beta}{\sin\nu} +$$
$$+ \sin\nu\sin\beta(\sin\beta\cos\gamma - \cos\xi\sin\gamma\cos\beta)\, ln\frac{1-\cos\nu}{1+\cos\beta} +$$
$$+ \phi(\xi, \pi-\alpha, \pi-\gamma, \lambda)\sin\xi\sin\gamma\sin\nu\sin\beta\Big), \tag{5}$$

$$q_{\alpha\beta}(\alpha, \beta) = \frac{\sin\nu}{\sin\beta}\ln\left(\frac{\alpha}{2}\frac{\nu}{2}\right) + \frac{\sin\nu}{\sin\alpha}\ln\left(\frac{\beta}{2}\frac{\nu}{2}\right) + \ln\left(\frac{\alpha}{2}\frac{\beta}{2}\right),$$

where $\phi$ is equal to

$$\phi(\xi, \alpha, \gamma, \lambda) = 2\arctan(\sin\xi \sin\alpha \sin\gamma\,, 1 - \cos\alpha + \cos\gamma + \cos\lambda).$$

The above formulae are applicable for arbitrary mutual arrangement of triangular panels $K_i$ and $K_j$, except some special case: if they lie in the same plane ($\xi = 0$) and one of the conditions $\beta = \gamma$ or $\alpha = \delta$ is satisfied, in the above formulae for $q_\Theta$ and $q_\Psi$ an ambiguity of the form [0/0] arises. Note that the same ambiguity also arises at $|\xi| \approx \pi$, however, let us do not consider this case, since it would correspond to an infinitely thin surface.

This problem can be solved by replacing the terms with ambiguities with their limit values, however, such approach in the general case does not allow for high accuracy achieving in calculating integrals (if one has in mind computations with double precision, typical for engineering practice, then it seems consistent to consider 'high accuracy' mean a relative error at the level not higher than $10^{-8}$). The latter is due to the fact that in practice, instead of the conditions $\xi = \gamma - \beta = 0$ or $\xi = \delta - \alpha = 0$, it is necessary to check 'weaker' conditions

$$|\xi| < \varepsilon \wedge |\gamma - \beta| < \varepsilon \quad \text{or} \quad |\xi| < \varepsilon \wedge |\delta - \alpha| < \varepsilon\,.$$

As a small parameter $\varepsilon$, it is impossible to take a value that is too small (for example, close to machine epsilon, about $10^{-14}$ or $10^{-15}$): in this case, even if the above 'criterion' for compliance with a special case is not met, in calculations being performed according to general formulae, eqns (2)–(5), problems of error accumulation will arise when subtracting close numbers. A numerical experiment shows that normally suitable are values no less than $\varepsilon = 10^{-6}$. Thus, the error of replacing of true values with the limit ones will be $O(\varepsilon)$, which, taking into account the above considerations, does not guarantee high accuracy in calculating the corresponding integrals.

This issue can be overcome by replacing of the expressions containing ambiguities not with their limit values [12], but with linear expansions, so that the resulting error is $O(\varepsilon^2)$. Expanding the expressions (4) and (5) into series with respect to small parameters up to linear terms and calculating the limit values of the coefficients, one can finally obtain

$$q_\Theta(\alpha, \beta, \mu, \gamma, \lambda) = \frac{\xi}{4}\left(5 + \cos 2\beta - 2\frac{\sin\frac{(\alpha-\beta)}{2}}{\cos\frac{\nu}{2}} - 2\frac{\sin^2\beta \sin\nu}{\sin\alpha}\ln\left(\tan\frac{\nu}{2}\tan\frac{\beta}{2}\right) - \right.$$
$$\left. \sin 2\beta \tan\frac{\alpha}{2}\right),$$

$$q_\Psi(\alpha, \beta, \mu, \gamma, \lambda) = \frac{1}{2} - \ln 2 - 2\ln\left(\sin\frac{\beta}{2}\right) + \frac{\sin\beta - \sin\nu}{\sin\alpha} + \frac{\cos\nu \sin\beta}{\sin\alpha}\ln\left(\tan\frac{\beta}{2}\tan\frac{\nu}{2}\right) +$$
$$+\frac{\gamma-\beta}{24\sin\alpha\sin\beta}\left(\sin\nu\left((3\cos 2\beta - 1)\ln(1 + \cos\nu) + 4\ln\frac{\sin\beta}{\sin\nu}\right) - 24\cos^2\frac{\beta}{2}\cos\nu\tan\frac{\nu}{2} + \right.$$
$$\left. +\sin\nu\left(4\ln\frac{1-\cos\nu}{1+\cos\beta} - (3\cos 2\beta - 1)\ln\left(2\tan^2\frac{\beta}{2}\sin^2\frac{\nu}{2}\right)\right)\right).$$

### 3.2 Panels with common vertex

Now let us consider the control panel $K_i$ and influencing panel $K_j$ with common vertex (Fig. 3). All auxiliary vectors and angles are also introduced in Fig. 3.

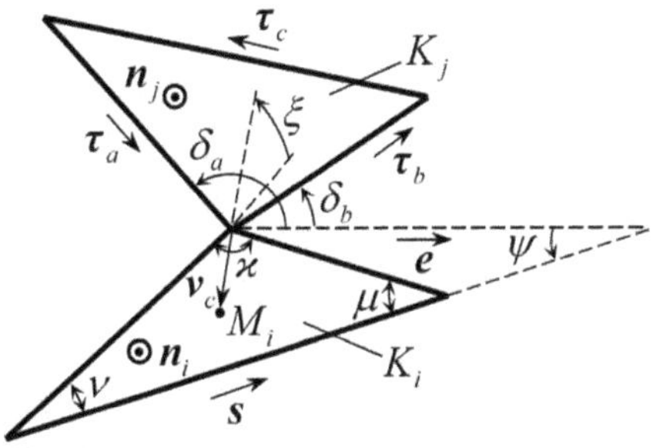


Figure 3: Two panels with common edge.

The unit direction vector for the line of intersection of the planes of the panels, denoted as $\boldsymbol{e}$, is chosen codirected to the vector product of the normal vectors to the panels $\mathbf{n}_i \times \mathbf{n}_j$. If the panels $K_i$ and $K_j$ lie in the same plane, i.e., at $\mathbf{n}_i \times \mathbf{n}_j = \mathbf{0}$ (in practice, for $|\mathbf{n}_i \times \mathbf{n}_j| < \varepsilon^2$), the direction of the unit vector $\mathbf{e}$ can be chosen arbitrarily; for clarity, let us take $\mathbf{e} = \boldsymbol{\tau}_b$. After introducing the unit vector $\mathbf{e}$, the orientation of the panel $K_j$ is fully determined by the angles $\delta_a$ and $\delta_b$ between its sides and the vector $\mathbf{e}$ (Fig. 3):

$$\delta_a = \arctan(-\mathbf{e}\boldsymbol{\tau}_a\mathbf{n}_j, -\mathbf{e}\cdot\boldsymbol{\tau}_a), \qquad \delta_b = \arctan(\mathbf{e}\boldsymbol{\tau}_b\mathbf{n}_j, \mathbf{e}\cdot\boldsymbol{\tau}_b).$$

If at least one of the angles $\delta_a$ or $\delta_b$ turns out to be close to $\pm\pi$, that in practice means checking the condition

$$\pi - |\delta_a| < \varepsilon \quad \vee \quad \pi - |\delta_b| < \varepsilon,$$

or, what is the same,

$$1 + \cos\delta_a < \frac{\varepsilon^2}{2} \quad \vee \quad 1 + \cos\delta_b < \frac{\varepsilon^2}{2},$$

it is convenient to change the direction of the unit vector $\mathbf{e}$ to the opposite, in this case the angle will become close to zero; the value of the second angle should be recalculated. The same operation should be performed when the following condition is met:

$$\delta_a \cdot \delta_b < 0 \quad \wedge \quad |\delta_a - \delta_b| > \pi.$$

After the direction of the unit vector $\boldsymbol{e}$ is finally determined, the angle $\xi$ between the planes of triangular panels $K_i$ and $K_j$ can be calculated:

$$\xi = \arctan(\mathbf{n}_i\mathbf{n}_j\mathbf{e},\ \mathbf{n}_i \cdot \mathbf{n}_j).$$

Positive or negative sign of the angle $\xi$ determines mutual orientation of the panels.

Then the expressions for the singular parts of the integral $\mathbf{J}(M_i, K_j)$ can be written in the following form:

$$\Theta^{\text{sing}}(M_i, K_j) = 2\left(\arctan\left(\mathbf{v}'_c\boldsymbol{\tau}_a\mathbf{e},\ (\mathbf{e}-\mathbf{v}'_c)\cdot(\mathbf{e}-\boldsymbol{\tau}_a)\right) + \arctan\left(\mathbf{v}'_c\boldsymbol{\tau}_b\mathbf{e},\ (\mathbf{e}-\mathbf{v}'_c)\cdot(\mathbf{e}+\boldsymbol{\tau}_b)\right)\right),$$

$$\boldsymbol{\Psi}^{\text{sing}}(M_i, K_j) = -\left(\boldsymbol{\tau}_a \ln\left(\frac{|\mathbf{v}_c|}{\sqrt{S_i}}(1+\boldsymbol{\tau}_a\cdot\mathbf{v}'_c)\right) + \boldsymbol{\tau}_b \ln\left(\frac{|\mathbf{v}_c|}{\sqrt{S_i}}(1-\boldsymbol{\tau}_b\cdot\mathbf{v}'_c)\right)\right).$$

Here, as earlier, $S_i$ means the area of the panel $K_i$; the prime means the unit vector of the corresponding vector.

As a result, the singular part $\Theta^{\text{sing}}$ and the multipliers for the vectors $\boldsymbol{\tau}_a$, $\boldsymbol{\tau}_b$ in the expression for $\boldsymbol{\Psi}^{\text{sing}}$ can be integrated analytically over the control panel $K_i$:

$$\int_{K_i} \Theta^{\text{sing}}(M_i, K_j)dS_r = S_i(q^{\Theta}(\delta_a) - q^{\Theta}(\delta_b) + 4\pi p), \tag{6}$$

$$\int_{K_i} \boldsymbol{\Psi}^{\text{sing}}(M_i, K_j)dS_r = S_i(q^{\Psi}(\delta_a)\boldsymbol{\tau}_a + q^{\Psi}(\delta_b)\boldsymbol{\tau}_b), \tag{7}$$

where $p$ is an integer, the meaning of which will be explained below; expressions for the functions $q^{\Theta}$ and $q^{\Psi}$ are the following:

$$q^{\Theta}(\delta) = \frac{2}{\sin\grave{u}\ \sin\psi}\left( A_\nu \sin\mu \sin(\nu+\psi) - A_\mu \sin\nu \sin(\mu-\psi) - \right.$$
$$\left. -\frac{1}{D}\sin\mu \sin\nu \sin\delta\left( W\cos\eta + \frac{\sin\psi \sin\xi}{2}(\Lambda_1 - \Lambda_2\cos\sigma)\right)\right), \tag{8}$$

$$q^{\Psi}(\delta) = \frac{3-\ln 2}{2} + \frac{\sin\mu \sin\nu}{\sin\grave{u}}\left( \ln\frac{1+\cos\lambda}{1+\cos\theta}\cot\psi + \right.$$
$$\left. +\frac{1}{D}\left( \Lambda_1\frac{\sin\delta\cos\eta}{\sin\psi} + \Lambda_2\cos\chi - 2W\sin\delta\sin\xi - G\ln\frac{\sin\nu}{\sin\mu}\right)\right) -$$
$$-\frac{1}{\sin\grave{u}}\left( \sin\mu\cos\nu\ln\frac{1+\cos\theta}{\sin\nu} + \cos\mu\sin\nu\ln\frac{1+\cos\lambda}{\sin\mu}\right) - \frac{1}{2}\ln\frac{\sin\mu\sin\nu}{\sin\grave{u}}. \tag{9}$$

Additional notations introduced for simplicity are the following:

$$A_\mu = \arctan\left( \tan\frac{\delta}{2}\cos\frac{\mu-\psi}{2}\sin\xi,\ \tan\frac{\delta}{2}\cos\frac{\mu-\psi}{2}\cos\xi + \sin\frac{\mu-\psi}{2}\right),$$

$$A_\nu = \arctan\left( \tan\frac{\delta}{2}\sin\frac{\nu+\psi}{2}\sin\xi,\ \tan\frac{\delta}{2}\sin\frac{\nu+\psi}{2}\cos\xi + \cos\frac{\nu+\psi}{2}\right),$$

$$W = \arctan\left( \sin\delta\sin\frac{\grave{u}}{2}\sin\xi,\ \cos\frac{\grave{u}}{2} + \sin\delta\cos\left(\frac{\mu-\psi}{2} - \frac{\nu+\psi}{2}\right)\cos\xi + \right.$$
$$\left. +\cos\delta\sin\left(\frac{\mu-\psi}{2} - \frac{\nu+\psi}{2}\right)\right),$$

$$D = \sin^2(\delta-\psi) + \sin\delta\sin\psi(1-\cos\xi)\big(\cos(\delta-\psi) + \cos\sigma\big),$$

$$G = \cos\psi\left( \sin\delta\cos\sigma\cos\xi + \cos\delta\cos\chi - \frac{\sin^2\delta}{\sin\psi}\right),$$

$$\Lambda_1 = \ln\left(\frac{1+\cos\lambda}{1+\cos\theta}\frac{\sin\nu}{\sin\mu}\right),\quad \Lambda_2 = \ln\left( \tan\frac{\nu}{2}\tan\frac{\mu}{2}\right).$$

Let us explain the meaning of the quantities in the presented formulae. The angle ù is the angle of the control panel $K_i$ at the vertex common with the influencing panel $K_j$; $\mu$ and $\nu$ are the other angles of the panel $K_i$ (Fig. 3); $\psi$ is the angle between the vector **e** and the directive vector **s** of the side of the panel $K_i$, which lies opposite the common vertex:

$$\psi = \arctan(\mathbf{e}\,\mathbf{s}\,\mathbf{n}_i,\ \mathbf{e}\cdot\mathbf{s});$$

for $\cos\sigma$, $\cos\chi$, $\cos\eta$, $\cos\theta$ and $\cos\lambda$ the following relations are valid:

$$\cos\sigma = \sin\delta\sin\psi\cos\xi + \cos\delta\cos\psi,$$

$$\cos\chi = \sin\delta\cos\psi\cos\xi - \cos\delta\sin\psi,$$

$$\cos\eta = \cos\delta\sin\psi\cos\xi - \sin\delta\cos\psi,$$

$$\cos\theta = \sin\delta\sin(\nu+\psi)\cos\xi + \cos\delta\cos(\nu+\psi),$$

$$\cos\lambda = \sin\delta\sin(\mu-\psi)\cos\xi - \cos\delta\cos(\mu-\psi).$$

The function $\Theta^{\text{sing}}(M_i, K_j)$ is given by the sum of two arctangents, so it is necessary to monitor the continuity of their branches by appropriate choice the value of the integer parameter $p$, however, the corresponding conditions are very cumbersome and inconvenient to check in practice. Instead, one can take into account the geometric sense of the function $\Theta(M_i, K_j)$: its absolute value is equal to the value of the solid angle at which the influencing panel $K_j$ is visible from the observation point $M_i$. Thus, its magnitude cannot exceed $2\pi$. Guided by this, after calculating the sum of two integrals

$$\int_{K_i}\Theta(M_i, K_j)dS_r = \int_{K_i}\Theta^{\text{sing}}dS_r + \int_{K_i}(\Theta - \Theta^{\text{sing}})dS_r,$$

where the first term is calculated using the presented analytical formulae, and the second numerically (since the integrand is a smooth function), it is necessary to choose a value of the parameter $p$ for which

$$-2\pi S_i \text{ „ } \int_{K_i}\Theta(M_i, K_j)dS_r \text{ „ } 2\pi S_i.$$

If the control and influencing panels do not have common internal points, this can be done in the only way.

The presented formulae for the integrals of $\Theta^{\text{sing}}(M_i, K_j)$and $\boldsymbol{\Psi}^{\text{sing}}(M_i, K_j)$ are valid for arbitrary relative positions of panels with a common vertex, except several special cases that arise for zero or close to zero values of $\sin\psi$, $\sin\delta$, $\sin\xi$ and some other quantities included in these expressions. According to the choice of the direction of the vector **e**, excluding the possibility of $\delta \approx \pm\pi$, one obtains that the condition $\sin\delta \approx 0$ is equivalent to the condition $\delta \approx 0$, while $\sin\psi$ and $\sin\xi$ vanish both at zero angles and at cases when the angles are equal to $\pm\pi$.

In the cases considered below, when using the given 'general' formulae, ambiguities of the type [0/0] or [$\infty - \infty$] arise. Let us present expressions for calculating the corresponding quantities, valid in the vicinity of these special cases and written in the form of expansions with respect to small parameter(s) with taking into account constant and linear terms. The upper asterisk index '*' indicates the limit value of the corresponding parameter.

1. Let the consider the case when $|\sin\xi| < \varepsilon$, $1 - |\cos\sigma| < \varepsilon^2/2$, and at the same time $|\sin\psi| \geq \varepsilon$. The small quantities here are $(\xi - \xi^*)$ and $(\delta - \cos(\xi^*)\psi - \Delta^*)$, where $\xi^*$ and $\Delta^*$ can take values 0 or $\pm\pi$. Calculating all the necessary limits, one obtains:

$$q^{\Theta}(\delta) = \frac{\pi}{2}(1-\cos\sigma)\left(1-\operatorname{sign}(\delta^*)\cos\xi^*\right)+$$

$$+(\xi-\xi^*)\left(1-\cos\sigma^*\left(\cos\psi+\frac{1}{2}\sin\psi(\cot\nu-\cot\mu)\right)+C\frac{\sin\psi}{2\sin\grave{u}}\right),$$

$$q^{\Psi}(\delta) = \frac{1}{2}\left(1 - ln\frac{2\sin\mu\sin\nu}{\sin\quad\varkappa}\right) +$$

$$+\frac{\cos\sigma^*}{\sin\grave{u}}\left(\left(1+\cos\nu\ln\left(\tan\frac{\nu}{2}\right)\right)\sin\mu-\left(1+\cos\mu\ln\left(\tan\frac{\mu}{2}\right)\right)\sin\nu\right)+$$

$$+\frac{1}{2}\left(\cos\xi^*(\delta-\Delta^*)-\psi\right)\left(\cot\nu-\cot\mu-C\frac{\cos\sigma^*}{\sin\grave{u}}\right),$$

where $C = \cot\mu\sin\nu + \cot\nu\sin\mu - \sin\mu\sin\nu\ln\left(\tan\frac{\mu}{2}\tan\frac{\nu}{2}\right)$.

2. If $|\sin\xi| < \varepsilon$, but at the same time $1 - |\cos\sigma| \geq \frac{\varepsilon^2}{2}$ and $|\sin\psi| \geq \varepsilon$, then one small parameter arises, namely $(\xi - \xi^*)$, where, as earlier, $\xi^*$ can take the values 0 and $\pm\pi$. The expansion of the function $q^{\Theta}$, which contains ambiguity, has the form

$$q^{\Theta}(\delta) = \frac{2}{\sin\grave{u}\,\sin\psi}\left(\arg\left(\operatorname{sign}(c_\varsigma s_\varsigma)\right)\frac{\sin\left((\cos\xi^*)\delta\right)\sin\mu\sin\nu}{\sin\varsigma}-\right.$$

$$\left.-\arg(\operatorname{sign}s_\varsigma)\sin\nu\sin(\mu-\psi)+\arg(\operatorname{sign}c_\varsigma)\sin\mu\sin(\nu+\psi)\right)+$$

$$+2(\xi-\xi^*)\frac{\cos\xi^*}{\sin\grave{u}\,\sin\psi}\sin\frac{\delta}{2}\left(\frac{\sin\mu\sin(\nu+\psi)}{c_\varsigma}\sin\frac{\nu+\psi}{2}-\frac{\sin\nu\sin(\mu-\psi)}{s_\varsigma}\cos\frac{\mu-\psi}{2}+\right.$$

$$\left.+\frac{\sin\mu\sin\nu}{\sin\varsigma}\cos\frac{\delta}{2}\left(\frac{\cos\xi^*\sin\delta}{c_\varsigma s_\varsigma}\cos\frac{\grave{u}}{2}-\frac{\sin\psi}{\sin\varsigma}\left(\ln\frac{s_\varsigma^2\sin\nu}{c_\varsigma^2\sin\mu}-\cos\varsigma\ln\left(\tan\frac{\mu}{2}\tan\frac{\nu}{2}\right)\right)\right)\right).$$

Here, for brevity, the notations are used:

$$\varsigma = (\cos\xi^*)\delta-\psi\,,\quad c_\varsigma = \cos\frac{\varsigma-\nu}{2},\quad s_\varsigma = \sin\frac{\varsigma+\mu}{2};$$

the function $\arg(z) = \arctan(\operatorname{Im} z, \operatorname{Re} z)$ means the principal value of the argument of the corresponding complex number.
Note that the function $q^{\Psi}(\delta)$ in this case does not contain ambiguity and can be calculated using the general formula, eqn (9).

3. If $|\sin\psi| < \varepsilon$, but $|\sin\delta| \geq \varepsilon$, then the expansions with respect to the small parameter $(\psi - \psi^*)$, where $\psi^* = 0$ or $\psi^* = \pm\pi$, up to linear terms, have the form

$$q^{\Theta}(\delta) = -\frac{\pi}{2}(1-\cos\xi)\left(1-\operatorname{sign}\psi^*\right)+\frac{2}{\sin\grave{u}}\left(T_\nu\sin\mu\cos\nu+T_\mu\sin\nu\cos\mu+\right.$$

$$+\frac{\sin\mu\sin\nu}{2\sin\delta}\left(2T\cos\delta\cos\xi+(L_f+L_p\cos\delta)\cos\psi^*\sin\xi\right)\Bigg)+$$

$$+\left(\psi-\psi^*\right)\left(\frac{\sin\xi}{\sin\delta}\left(\cos\delta\frac{\sin\nu-\sin\mu}{\sin\text{ù}}+1\right)+\right.$$

$$\left.+\frac{\sin\mu\sin\nu}{\sin\text{ù}\,\sin^2\delta}\left(T\left((1+\cos^2\delta)\cos 2\xi+\sin^2\delta\right)+\cos\psi^*\sin 2\delta\left(L_f\cos\delta+L_p\frac{1+\cos^2\delta}{2}\right)\right)\right)\Bigg),$$

$$q^{\Psi}(\delta)=\frac{3}{2}-\frac{1}{2}\ln\frac{2\sin\mu\sin\nu}{\sin\text{ù}}-\frac{1}{\sin\text{ù}}\left(\cos\nu\sin\mu\ln\frac{A}{\sin\nu}+\sin\nu\cos\mu\ln\frac{B}{\sin\mu}-\right.$$

$$\left.-\frac{\sin\mu\sin\nu}{\sin\delta}\left((L_f\cos\delta+L_p)\cos\psi^*\cos\xi-2T\sin\xi\right)\right)+$$

$$+\left(\psi-\psi^*\right)\left(\frac{\cos\xi}{\sin\delta}\left(\cos\delta+\frac{\sin\nu-\sin\mu}{\sin\text{ù}}\right)+\right.$$

$$\left.+\frac{\sin\mu\sin\nu}{\sin\text{ù}\,\text{ssin}^2\delta}\left(L_f\left((1+\cos^2\delta)\cos 2\xi+\sin^2\delta\right)+\cos\delta\left(L_p\cos 2\xi\cos\psi^*-2T\sin 2\xi\right)\right)\right).$$

The following notations are used here:

$$A=1+\cos\delta\cos\nu+\cos\psi^*\cos\xi\sin\delta\sin\nu,\quad B=1-\cos\delta\cos\mu+\cos\psi^*\cos\xi\sin\delta\sin\mu,$$

$$T_\nu=\arctan\left(\sin\xi\sin\nu\tan\frac{\delta}{2},\ (1+\cos\nu)\cos\psi^*+\cos\xi\sin\nu\tan\frac{\delta}{2}\right),$$

$$T_\mu=\arctan\left(\sin\xi\sin\mu\tan\frac{\delta}{2},\ (1-\cos\mu)\cos\psi^*+\cos\xi\sin\mu\tan\frac{\delta}{2}\right),$$

$$T=\arctan(\sin\text{ù}\ \sin\delta\sin\xi,\ A+B+\cos\text{ù}\ +1),$$

$$L_f=\ln\frac{A\sin\mu}{B\sin\nu},\quad L_p=\ln\left(\tan\frac{\mu}{2}\tan\frac{\nu}{2}\right).$$

4. When $|\sin\psi|<\varepsilon$ and $|\sin\delta|<\varepsilon$ the expression $q^{\Theta}(\delta)$ depends only on the small parameter $(\psi-\psi^*)$, where, as earlier, $\psi^*=0$ or $\psi^*=\pm\pi$:

$$q^{\Theta}(\delta)=\frac{\pi}{2}\left(1-\cos\psi^*\right)(1+\operatorname{sign}\psi^*)+\left(\psi-\psi^*\right)S\sin\xi.$$

In this case, the expansion of $q^{\Psi}(\delta)$ also contains a term proportional to the small parameter $\delta$:

$$q^{\Psi}(\delta)=\frac{1}{2}\left(\operatorname{sign}\left(\cos\psi^*\right)\left(\ln\frac{\tan\nu/2}{\tan\mu/2}+\frac{\sin(\mu-\nu)}{\sin\text{ù}}\ln\left(\tan\frac{\mu}{2}\tan\frac{\nu}{2}\right)-2\frac{\sin\nu-\sin\mu}{\sin\text{ù}}\right)+\right.$$

$$\left.+\ln(\cot\nu+\cot\mu)+1-\ln 2\right)+\left(\delta-\left(\psi-\psi^*\right)\cos\xi\right)S,$$

where $S=\frac{\cos\psi^*}{2\sin\text{ù}}\left((\cos\mu+\cos\nu)\frac{\tan\nu/2}{\tan\mu/2}-\sin\mu\sin\nu\ln\left(\tan\frac{\mu}{2}\tan\frac{\nu}{2}\right)\right).$

## 4 CONCLUSION

In the present paper, analytical expressions convenient for use in practical computations are obtained for integrating the gradient of the fundamental solution of the Laplace equation over a triangular cell of a surface mesh. A semi-analytical technique for calculating repeated integrals of unbounded functions has been implemented, which made it possible to propose an approach for calculating repeated integrals of the gradient of the Newtonian potential over two cells. All possible cases of mutual arrangement of triangular cells are considered, including special cases when analytical formulas contain ambiguities. The source code of the implementation of the suggested algorithm of integration is freely available on github (https://github.com/vortexmethods/integrator).

## ACKNOWLEDGEMENTS

This paper has been supported by the Kazan Federal University Strategic Academic Leadership Program ('PRIORITY-2030').

## REFERENCES

[1] Brebbia, C.A., Telles, J.C.F. & Wrobel, L.C. (eds), *Boundary Element Techniques*, Springer-Verlag: Berlin and New York, 1984.

[2] Banerjee, P.K. & Butterfield, R., *Boundary Element Methods in Engineering Science*, McGraw-Hill: New York, 1981.

[3] Katsikadelis, J.T., *Boundary Elements: Theory and Applications*, Elsevier Science: Oxford, 2022.

[4] Cottet, G.-H. & Koumoutsakos, P.D., *Vortex Methods: Theory and Practice*, Cambridge University Press: Cambridge, 2000.

[5] Mimeau, C. & Mortazavi, I., A review of vortex methods and their applications: From creation to recent advances. *Fluids*, **6**, 68, 2021.

[6] Dergachev, S.A., Marchevsky, I.K. & Shcheglov, G.A., Flow simulation around 3D bodies by using Lagrangian vortex loops method with boundary condition satisfaction with respect to tangential velocity components. *Aerospace Science and Technology*, **94**, 105374, 2019.

[7] Reid, M.T.H., White, J.K. & Johnson, S.G., Generalized Taylor–Duffy method for efficient evaluation of Galerkin integrals in boundary element method computations. *IEEE Transactions on Antennas and Propagation*, **63**(1), pp. 195–209, 2015.

[8] Dodig, H., Cvetkovi, M. & Poljak, D., On the computation of singular integrals over triangular surfaces in $R^3$, *WIT Transactions on Engineering Sciences*, vol. 122, WIT Press: Southampton and Boston, pp. 95–105, 2019.

[9] Lifanov, I.K., Poltavskii, L.N. & Vainikko, G.M., *Hypersingular Integral Equations and their Applications*. Charman&Hall: Boca Raton, 2004.

[10] van Oosterom, A. & Strackee, J., The solid angle of a plane triangle. *IEEE Trans. on Biomedical Eng.*, **2**, pp. 125–126, 1983.

[11] Dunavant, D.A., High degree efficient symmetrical Gaussian quadrature rules for the triangle. *Int. J. for Numerical Methods in Engineering*, **21**(6), pp. 1129–1148, 1985.

[12] Marchevky, I. & Shcheglov, G, Numerical computation of double surface integrals over triangular cells for vortex sheet intensity reconstruction on body surface in 3D vortex methods. *WIT Transactions on Engineering Sciences*, vol. 122, WIT Press: Southampton and Boston, pp. 47–61, 2019.

# AUTOMATIC ROBOT CAR PARALLEL PARKING SYSTEM USING ARTIFICIAL NEURAL NETWORK

SARA GHATTA[1], SORAYA ZENHARI & AMIRHASSAN MONADJEMI[2]
[1]Institut für Software Engineering (ISTE), Germany
[2]Department of Computer Science School of Computing, National University of Singapore, Singapore

ABSTRACT
In today's busy world, parking a car in crowded cities is time-consuming and challenging. Parallel parking systems are a valuable innovation, especially for disabled and inexperienced drivers, as they can prevent collisions. Despite their common use, artificial neural networks (ANNs) and backpropagation algorithms (BP) have issues, such as difficulty in estimating network configurations and low accuracy. While BP is traditionally used to train ANNs, it is not effective for parking a car. To achieve accurate vehicle control, a self-organising map (SOM) clusters data from various parking scenarios and uses this information to classify the data. According to simulations conducted using MATLAB, SOM provides better accuracy and reliability for parking cars compared to BP.
*Keywords: self-organising map algorithm, automated vehicle, car-like robot parallel parking, artificial neural network (ANN), robot navigation, data scaling.*

## 1 INTRODUCTION

In today's hectic and stressful life, parking a car accurately is a challenge for many people There are many problems associated with parallel parking in cities because of their crowded streets and limited parking spaces, which often leads to vehicle damage. This task is particularly difficult for new drivers and can contribute to traffic congestion. To address these problems, automatic car parallel parking systems have emerged, offering significant benefits, especially for disabled and inexperienced drivers [1]. Automatic Parking processes can be categorised into three parts, including environment perception, path planning, and tracking using a control strategy Perception of the environment defines a parking space as the space formed by two obstacles or lines painted on the road. For a parking space formed by obstacles, methods based on range sensors have been developed [2]. Ultrasonic sensors and light detection and ranging (LiDAR) have been used to determine the shape of each obstacle. Simultaneous localisation and mapping (SLAM) have been established during the parking process, to locate the vehicle at the cm level. Path planning is considered in finding a collision-free geometric path from an initial position to a final position while traveling in the route. Trajectory planning is in connection with real-time planning, which is parameterised by time as well as velocity and acceleration [3]. It can exist many feasible paths, but we often require the shortest path in length.

### 1.1 Background and preliminaries

Automatic path planning and path tracking technology have attracted many researchers in recent years. Parking path planning approaches are mainly characterised into four sets: geometric curve approaches [4], graph search approaches [5], random sampling approaches [6], and intelligent optimisation approaches [7]. In the geometric curve technique, Vorobieva et al. [8] examined tiny parking places characterised by the clothoid curve and considered parking pathways appropriate for tiny parking places. But the clothoid curvature design was moderately complex demanding a large number of mathematical operation and calculation. In the graph search technique, Ren et al. [5] planned an automatic parking path planning

WIT Transactions on Engineering Sciences, Vol 136, © 2024 WIT Press
www.witpress.com, ISSN 1743-3533 (on-line)
doi:10.2495/BE470051

system applying the Hybrid A* procedure and RS curves, and then optimised the pathway by means of Bezier curves and gradient descent method. The study enhanced path search skill and flexibility in complex situations and resolved the disjointed curvature's issues, but the pathway was not sufficiently flat and it causes weakness in the tracking result. In the random sampling technique, Zhang et al. [6] proposed the series of principles for taking the reference points and updating route nodes that matched to the kinematic limitations, enhancing the discovering random tree algorithm, speedily. The results exposed that the new process was able to carefully control the vehicle to perform the parking works, but the tracking was problematic. In the intelligent optimisation technique, Chen et al. [9] suggested a training way utilising the syllabus learning with the aid of the position of rear axle's centre and the steering angle of front wheel as state variables, reached an arrangement success percentage of 91% though sufficient comfort based on deep reinforcement learning. Yet, it's process needed large numbers of data.

Presently, furthermost studies on path planning for automatic parking can mollify the curving limitations and difficulty avoidance limits throughout the process. Current parking route tracking approaches contain pure pursuit (PP) method [10], model predictive control (MPC) [11], and fuzzy control method [12]. Horváth et al. [13] recommended pure pursuit method. The PP algorithm modified lateral deviation by picking diverse target points and dynamically adjusting the preview interval. Experimental confirmation displayed good viability in real-world applications and improved tracking routine. Wu et al. [11] recommended a parking pathway pursuing control system applying MPC, which tracked routes involving clothoid curves, semicircles, and straight lines. The results showed that the method had excellent precision. Nakrani and Joshi [14] established a parallel parking process applying hybrid fuzzy inference model, which figured out several parking circumstances, counting sudden difficulty entrance throughout parking, the experimental results indicated that this technique could effectively complete parking in various parking situations, but its correctness was moderately weak.

Daxwanger and Schmidt [15] investigated path planning and tracking algorithms for automatic parallel parking, combining neural networks and fuzzy logic to achieve automatic parallel parking. Scicluna et al. [16] proposed a hardware solution for vehicle parallel parking and regarded fuzzy logic as a fast approach that could be implemented with the hardware. Cheng et al. [17] proposed fault-based steering control algorithm for segmental path planning problem for automatic parallel parking assist system in narrow parking places. Vorobieva [18] investigated automatic parallel parking with geometric continuous-curvature path planning. A geometric path was created by some arc, and it was transformed to the continuous path. Likewise, automatic parallel parking in a tiny spot, utilising path planning and control was reported by Vorobieva et al. [19]. Naderi Samani et al. [20] suggested a p-domain path-tracking controller for parking a car like mobile robot (CLMR). A path planning approach composed of two circular arcs connected by a tangential point was offered by Ji et al. [21] based on preview BP neural network PID controller. Li et al. [22] considered interior-point method (IPM) based on the simultaneous dynamic optimisation approach to solve the problem of automatic vehicle parallel parking. This global optimum method took a long time to find the optimal parking manoeuvre. Wu et al. [23] designed an ultrasonic sensor and applied the smart wheeled mobile robot (SWMR) automatic algorithm to solve car parallel parking problem depending on the experience of the actual drives. Nedamani et al. [24] proposed a fifth-degree polynomial curve for autonomous parallel parking, based on interpolating algorithms. The proposed curve offers the advantages of low computational cost, comfort, and independence from global waypoints.

Many path planning and control methods have been proposed to produce the parking function. Sampling-based strategies (SBS) have largely been used in robotic applications. In some SBS, planning is based on randomly sampling configuration space and finding connectivity inside this space. Common algorithms of SBS, which is also applied in parking, are rapidly-exploring random tree (RRT) [25], Hybrid A* search [26] and optimisation-based path planning strategy [27]. RRT methods [28] can plot in continuous space using heuristic information. The A* search method [29] discretises the search space, which guaranteed completeness and optimality. However, choosing the mesh size requires specialised knowledge. In the optimisation methods, local optimisation [30] was developed to find the shortest path in a short time; although global optimality cannot be guaranteed.

For marching methods, linear quadratic regulator (LQR) [31], sliding function control [32] and Fuzzy controller [33] were used. Fuzzy Logic based methods have been applied in the numerous research. Chen and Feng [34] combined both fuzzy inference and self-organising map (SOM) neural networks to approach the intelligent vision-based car-like vehicle backing system. This study not only explains the computer simulation's results, but also proves the practical car parking achievements in a real situation. Joshi and Zaveri [35] proposed a neuro-fuzzy system for navigating mobile robots; it contained finding distances from an obstacle and training how to control movement with neural network. Marasigan et al. [36] assumed Neuro-fuzzy controller for data which gained from a fifth-degree polynomial path in backward manoeuvre. Utilising the genetic algorithm, Aye et al. [37] optimised the image-based fuzzy controller for an automatic parking system, based on digital image processing. With the genetic algorithm the system can be flexible to parking conditions. Nari et al.[38] designed an automatic parking system using trajectory control method with face-to-face feedback. They assumed that with the addition of a gyroscope sensor, the system can know the direction facing the actual car and adjust the direction towards the destination using fuzzy logic control. The use of gyro sensors and fuzzy logic control in automatic parking systems greatly affects parking accuracy.

ANN approaches have been widely considered in the planning problems for parking. Heinen et al. [39] published an article on autonomous vehicle parking and pull out using ANNs; He designed a vehicle which could recognise environment by a sensor and used the neural networks to generate steering angle and velocity.

Farooq et al. [40] implemented a mobile robot navigation with ANNs using BP algorithm to solve the navigation problem. They mentioned that neural network has been a superior method which can learn complex nonlinear relationship between inputs and outputs. But applying the tangent sigmoid activation function would have made errors in the system and caused less efficiency in the neural controller. Considering the steering angle and speed as outputs, Zhou [41] studied on automatic car parallel parking and reviewed various methods to convert human expertise to machine learning in numerous iterations by ANNs.

Ji et al. [21] investigated path planning and tracking for vehicle parallel parking based on BP Neural Network PID Controller, experimentally. Liu et al. [42] presented a method to enumerate all the possible parking trajectories and corresponding steering actions, and then have the parking controller learn the relationship between the given initial-and-final state pairs and the corresponding sequence of steering actions using an ANN. Li et al. [43] proposed an end-to-end neural-network-based automatic parking controller. Moon et al. developed an automatic parking controller with a twin ANN architecture [44]. Ma and Wang [45] Mediate perception utilised AI techniques like convolutional neural networks (CNNs) to identify one or more objects. A well-established perception task effectively handled by AI is traffic sign recognition. The accuracy of AI such as deep neural networks (DNNs) has achieved 99.46. Genetic algorithm is used for detecting lanes for the navigation of AVs.

Kockelman proposed CNNs for detect the barrier with the accuracy of 80%. The SVM is proposed for more classification of CNN learnt feature. LSTM is used to learn camera images in order to determine for the wheel's angels. The result shows the LSTM outperform the CNN in this scenario. In total, the perception and situation awareness has a better contribution in the autonomous driving. The neural Network has suffered from the false detection. Although CNN, LSTM, and DBN for autonomous vehicle has proposed, it has still some issues. The DL use more intricate model structures, making parameter calibration computationally costly. Lack of the model for the hidden layers for training is an issue, and user has to do it within some trial and error. Previous optimisation-based approaches can generate accurate parking trajectories, but these methods cannot compute feasible solutions with extremely complex constraints in a limited time. Besides, the control methods, cannot address system limitations such as steering actuation limits. Recent research uses ANN-based approaches that can generate time-optimised parking trajectories in faster time without limitation. Recently, methods based on deep neural networks have been expected to solve the drawbacks of other offline approaches by manoeuvring vehicles without prior offline trajectory planning. ANNs consist of interconnected neurons that process information collaboratively to solve problems. By training an ANN using a dataset generated by simulation or experiment, the ANN learns hyper-dimensional relationships between the current vehicle states and the appropriate vehicle manoeuvring signals. Instead of calculating the parking trajectory offline, the ANN-based parking controller can yield a direct manoeuvring signal of the steering angle and velocity online, while the vehicle is moving into a parking space.

As mentioned before, a common training algorithm previously used for ANNs for collision avoidance in robot motion was BP. It is a supervised method, but this algorithm had some drawbacks; It was very complicated and slow in convergence. This algorithm has problem in estimation, there was no defined amount of hidden layer for the network, and it was arduous to estimate the number of hidden layers in this network. These are some significant issues to make this algorithm inefficient for using in car parking project. Generally, parking a car-like robot with this algorithm was not only imprecise but also car might be collided with obstacles in some circumstances. Hence, the present study aims to improve automatic car parallel parking systems using ANN with SOM algorithm. It explores an AI-based method to control a car-like robot, focusing on automatic parking in various circumstances. Discussing about the disadvantages of BP algorithm, this paper aims to achieve a more efficient collision-free trajectory planning scheme for automated parking in narrow parking spaces.

## 2 RESEARCH METHOD

### 2.1 Self-organising map definition

To date, many methods have been proposed for parallel parking cars. However, this article presents an unsupervised algorithm based on artificial neural networks (ANN) called the self-organising map (SOM) [46]. This algorithm clusters similar datasets together by comparing input data with neuron links. When a similarity is detected between the input data and the neurons, the neurons adjust to become even closer to the inputs. Consequently, neighbouring neurons collaborate with this primary neuron and also adjust to get closer to the input data. In this way, all data are effectively clustered.

## 2.2 Dataset preparation

As shown in Fig. 1, it is crucial to choose a suitable and appropriate robot for simulation, as a robot's degree of manoeuvrability, including both mobility and steerability, differs for each type. For this simulation, an Ackerman robot is employed.

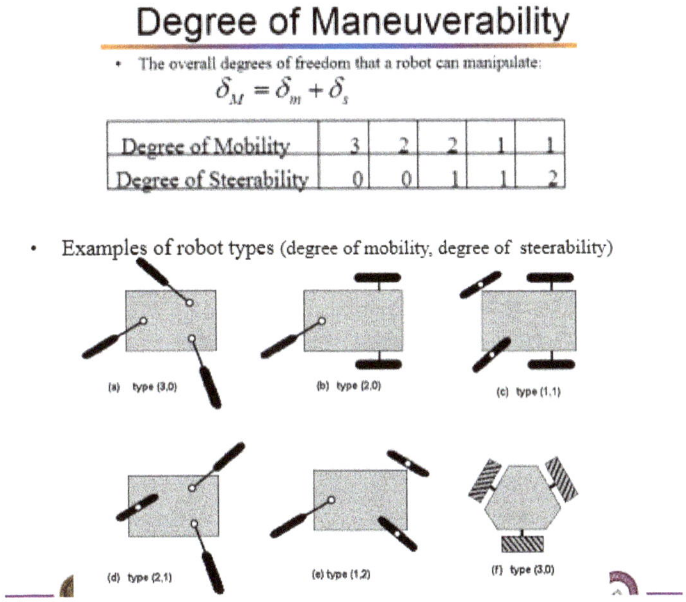


Figure 1: Different types of robots based on their degree of manoeuvrability.

In order to prepare adequate data, a displacement sensor capable of measuring distances between the four sides of the car-like robot and potential obstacles in various movements and scenarios will be utilised. Data from these sources are used to create the feature vector, which is considered as input data by the robot controller. In addition, steering angle and velocity are presumed to be outputs of the robot programs. After obtaining the distance between the four sides of the Ackerman robot and the obstacles during each movement, the data is classified based on the steering angle and speed of the robot within each class. As another feature vector, the state of movement variable is considered. As shown in Fig. 2, the steering angle and direction of the vehicle play an important role in the classification process.

Once all distances between the four sides of the car-like robot and any possible obstacles have been measured, the state of movement is acquired according to both steering angle and velocity. According to Fig. 3, both the state of movement and all distances gained from multiple scenarios are taken into consideration as inputs to the neural network. By training these input data with SOM algorithm, similar input data would be categorised by SOM. It should be noted that SOM determined multiple clusters for these similar input data, and these data will be replaced by a single cluster in datasets. The remaining elements in the dataset are clusters, steering angle, and velocity. Therefore, the clusters are designated as inputs, while the velocity and steering angle are designated as outputs. The SOM training process was conducted using the NCTOOL in MATLAB.

WIT Transactions on Engineering Sciences, Vol 136, © 2024 WIT Press
www.witpress.com, ISSN 1743-3533 (on-line)

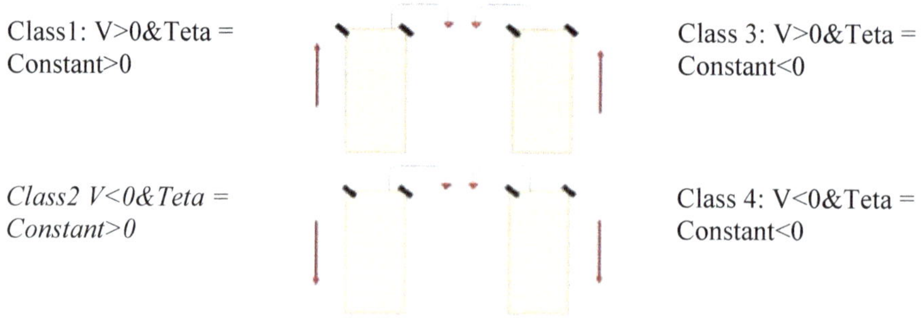


Figure 2: The state of movement according to the state of both velocity and steering angle.

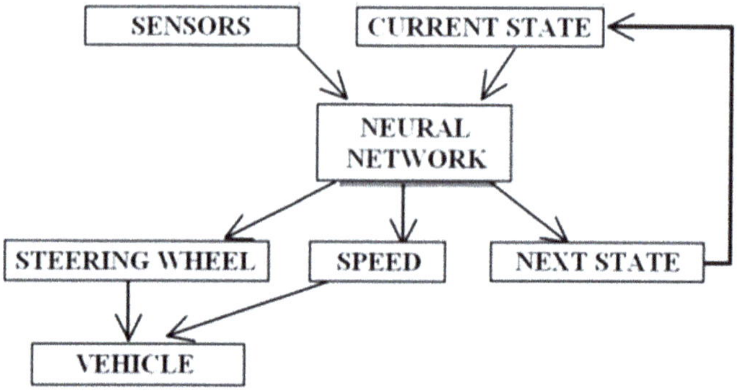


Figure 3: Artificial neural networks model scheme.

The experiments have resulted in various steering angles, but each cluster maintains a consistent speed. Therefore, it is optimal to consider the mode of the steering angles within each cluster. For each cluster, the input data consist of a consistent speed and the mode of the steering angles as outputs. Following the training phase, it is crucial to evaluate the car's performance by testing the trained neural network. Similar to the training phase, input data are processed by the SOM in the test phase, but now the input data are obtained from a single parking scenario. This generates a list of clusters specific to that scenario. Finally, the steering angle and speed associated with each cluster, obtained from the training phase across multiple scenarios, are applied to the corresponding cluster in the current parking scenario. These parameters – steering angle and speed – are essential for guiding the car-like robot simulator through navigation tasks facilitated by the ANN.

## 3 IMPLEMENTATION

Once a car moves forward or backward, it undergoes a rotation, resulting in a change in its orientation angle. If a car moves from an initial point to a second point, the new orientation angle of the car is the sum of its initial angle and the rotation angle it undergoes. Following this, a kinematic model is essential to convert the local coordinates to the global coordinates, considering the updated orientation angle of the car. This model helps in accurately determining the position and orientation of the car relative to its movement and rotation.

### 3.1 Ackerman kinematic model needed for simulation

According to the kinematic model, it is necessary to convert local coordinates to global coordinates in order to use any simulation. In this model, the front wheels can be turned, while the back wheels cannot. This is similar to the Ackerman robot used for providing datasets. Fig. 4 demonstrates the kinematic model. In this figure, the car-like robot is represented by a rectangle with four wheels. In eqns (1) and (2), X and Y indicate the coordinates of the rear wheels, θ represents the rotation angle of the vehicle, $\varphi$ is the steering angle. Additionally, in eqn (3), L is the wheelbase. Furthermore, the car-like robot can be moved by two commands ($\varphi$, V). Eqns (1)–(3) were derived according to the nonholonomic motion estimation [47].

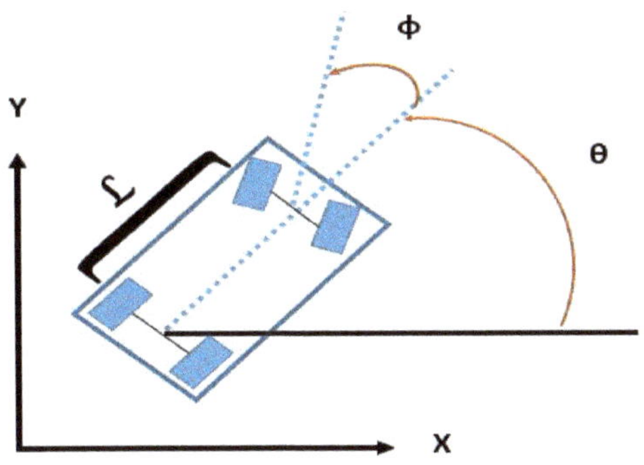


Figure 4: The kinematic model.

$$\dot{x} = v \cos \varphi \cos \theta \tag{1}$$

$$\dot{y} = v \cos \varphi \sin \theta \tag{2}$$

$$\dot{\theta} = \frac{v}{l} \sin \varphi \tag{3}$$

## 4 RESULTS AND DISCUSSION

### 4.1 Results of the back propagation algorithm employment

Although we presume all distances and states of movement as inputs to the neural networks, we cannot directly use them in the backpropagation network. Feature scaling or normalisation is essential for preparing data, including measurements of distances between the robot and possible obstacles, for efficient training. Without this step, we would have redundant neurons and low training accuracy. Eqn (4) is used for feature scaling. Similar to the SOM, input data in backpropagation must be classified. After training the input data with the backpropagation algorithm, we obtained a network with 5 inputs, 25 hidden neurons, and 2 outputs, as shown in Fig. 5. This process was carried out using MATLAB's NFTOOLS.

$$x' = \frac{a - \bar{x}}{\sigma} \tag{4}$$

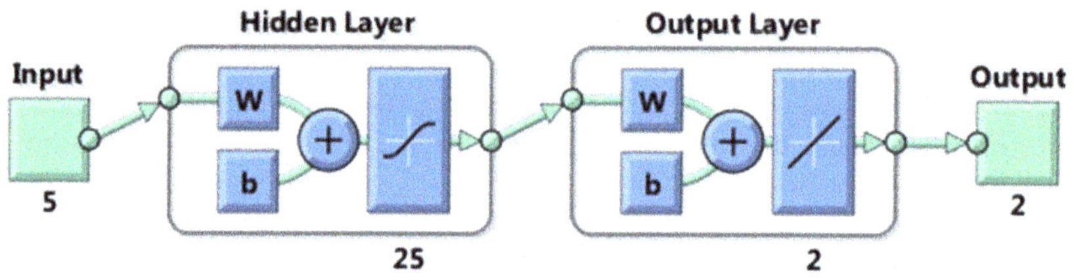


Figure 5: Topology of network after training data with the backpropagation algorithm.

After completing the training phase with multiple scenarios, the learned model will be applied to a different parking scenario during the test phase. Similar to the training process, scaling and classification of input data will be conducted. However, in this case, only data from a single scenario is available for testing. Once the input data are processed and the model is trained, the outputs will be speed and wheel angles. These outputs need to be converted from the scaled format used during training back to the original format suitable for the simulation program. This conversion is crucial because the simulation program operates based on these two parameters – speed and wheel angles – to accurately simulate the car's movements in the parking scenario. Therefore, maintaining accuracy and fidelity in this conversion process is essential for realistic simulation outcomes.

## 4.2 Analysing backpropagation with its plots

Fig. 6 illustrates the experimental results for backpropagation training. The figure shows the error plotted against epochs. Typically, the error decreases as training progresses through multiple epochs. Training is typically halted when the model begins to approach overfitting, which occurs when the model performs well on the training data but fails to generalise to unseen data. In this study, the best performance was achieved at 93 epochs with a mean squared error (MSE) of approximately 0.1. This point represents a balance where the model has learned effectively from the training data without overfitting, as indicated by continued improvement in validation error.

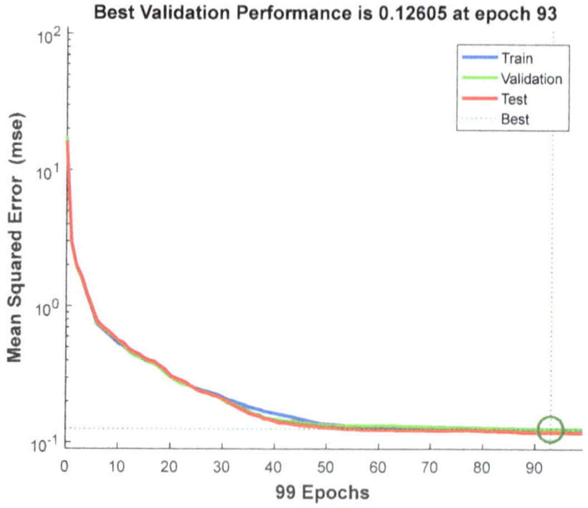


Figure 6: MSE versus epochs performance plot.

Fig. 7 depicts the outcomes of the training, validation, and test phases in a regression plot. This plot illustrates the relationship between output and target values, with the best outcome occurring when the output exactly matches the target. In regression analysis, an ideal scenario is represented by R=1, indicating perfect correlation between output and target values. In this experiment, the regression coefficient (R) is 0.93, indicating a strong correlation. A line of best fit, known as the fit line, is drawn to assess how closely it aligns with the dashed line representing perfect alignment. The closer the fit line matches the dashed line, the better the neural network performs, with lower error. Error in this context refers to the difference between the target (desired output) and the actual output. However, the scattered data points indicate variability in the data which makes it challenging for backpropagation to accurately predict outputs based on these inputs. As a result, achieving optimal performance through backpropagation can be difficult with such scattered data.

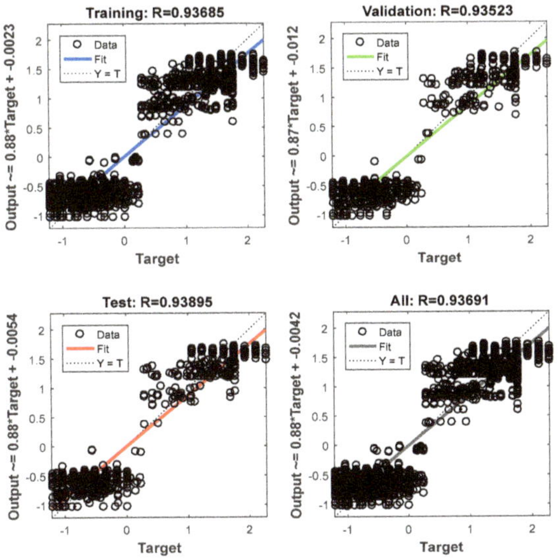


Figure 7: Regression plot includes the data position in three phases training, validation and test.

As shown in Fig. 8, the steering angle data in backpropagation exhibits significantly more scattering compared to that in the SOM. Backpropagation data is dispersed far from the central line, whereas SOM data clusters closely around it. This analysis underscores the distinct behaviour of data within these algorithms. In all charts referenced above, the vertical axis represents output data, while the horizontal axis represents target data. Both axes are divided into positive and negative areas due to the sign of the steering angle.

In Fig. 9(b), the actual (target) steering values range from +22 to +41, whereas the predicted steering values by backpropagation range from +25 to +59. This discrepancy indicates that backpropagation has lower precision in estimating positive angles. Conversely, in the negative angles, where actual steering values range from −26 to −58, backpropagation predicts values ranging from −25 to −58. Data clustering is more pronounced in the negative region, suggesting that backpropagation estimates are more accurate there compared to the

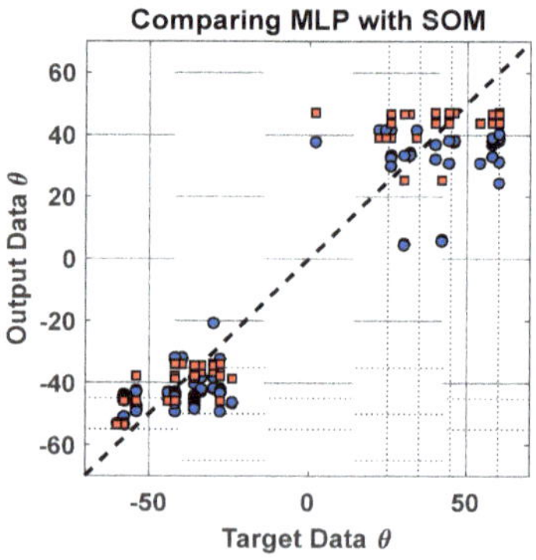


Figure 8: Comparison of data scattering in SOM with back propagation.

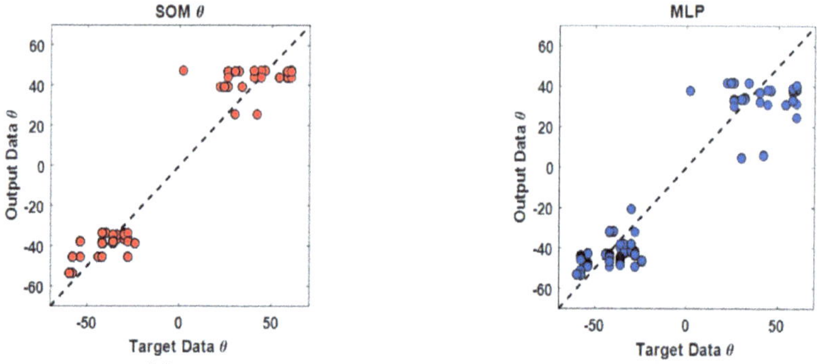


(a) (b)

Figure 9: Target steering angle data vs predicted steering angle data. These results were obtained after training input data by SOM labelled in (a) and after training input data by backpropagation labelled in (b).

positive region. According to Fig. 9(a), SOM predicts steering data ranging from +20 to +50 in positive angles, while the actual data ranges from +24 to +58. In the negative angles, where actual steering data ranges from −32 to −60, SOM predicts data ranging from −24 to −53. Importantly, despite the differences in value ranges, SOM exhibits significantly lower data scattering; the dispersion around the central line is minimal, indicating closer proximity to the desired outputs (black line). This highlights the effectiveness of SOM in this context. The superiority of the SOM over backpropagation in training car parallel parking data stems from its ability to cluster data more effectively and reduce data scattering, thereby achieving more accurate predictions and better performance overall.

In Fig. 10, the comparison between errors in the SOM and backpropagation is illustrated. As depicted in Fig. 10(a), the red square points are higher than the blue points, and the data around the central line is less scattered compared to the blue points. The red points align closely with the central data, indicating that SOM provides highly accurate estimates for these data points. Notably, the predicted data using SOM closely matches the target data. On the other hand, Fig. 10(b) shows that data points in backpropagation multi-layer perceptron (MLP) do not align precisely with the central line. There is a noticeable discrepancy between the target data and the predicted data in backpropagation. The target velocity data ranges from −6 to 6, whereas the predicted values range from 3 to 6 for positive velocity and from 4 to 8 for negative velocity. This variability suggests that backpropagation MLP struggles with accuracy in both velocity and steering angle predictions. Due to the classification approach utilised by SOM, it demonstrates greater precision in estimating speed. This precision stems from the fact that speed predictions are derived from well-defined classes within SOM. Consequently, SOM consistently delivers more desirable results compared to backpropagation MLP, making it the preferred method for car parallel parking tasks in terms of both velocity and steering angle estimation.

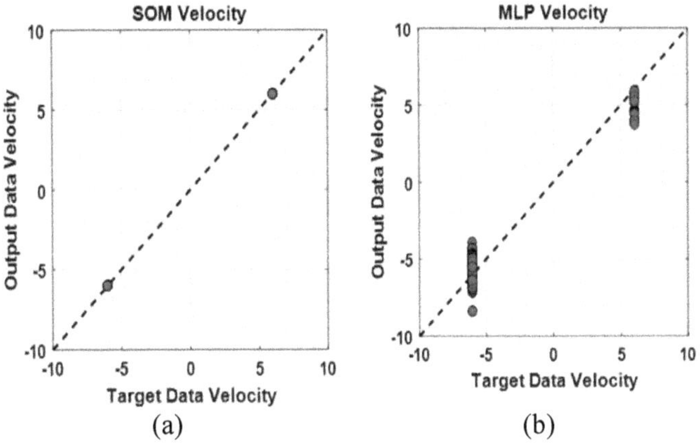


Figure 10: Comparison between target velocity data with predicted steering angle data in both SOM and backpropagation algorithm.

## 5 CONCLUSION

This article presents a survey on using neural networks for car parallel parking, comparing traditional methods like backpropagation with MLP structures to a proposed SOM algorithm. The study explores both the advantages and disadvantages of using backpropagation for this application. To improve car parallel parking, the SOM-based algorithm is introduced. A car-like robot was assembled and programmed based on datasets gathered from a distance sensor integrated into the robot. During the training phase, datasets from multiple scenarios were fed into the neural network to learn the relationship between input (sensor data) and output (steering angle and speed). This enabled the network to predict outputs for new inputs during the test phase. A simulation program was then developed based on the output dataset obtained from the test phase. Tools such as NetLab and RapidMiner were considered for neural network implementation, but practical experiments were conducted using MATLAB's NCTOOL. It was observed that dataset feature scaling improved vector balance, enhancing the performance of the experiments on the SOM. As part of the experiment, NCTOOL was used to classify the input dataset for a variety of scenarios. Outputs were clustered during both the training and test phases, where each cluster exhibited varied steering angles while maintaining a consistent speed established during training. In simulations involving numerous small movements, parking accuracy using the self-organising map outperformed backpropagation MLP. The density of data posed a challenge for backpropagation in establishing effective function relationships, whereas the self-organising map efficiently clustered data regardless of density, thereby controlling the vehicle parking procedure effectively. In the future, researchers may explore deep learning methodologies to improve the performance of automatic neural network-based controllers for parallel parking cars by leveraging additional sensors with higher precision and performance. With continuous advancements like this, vehicle control systems based on neural networks will continue to be refined.

## REFERENCES


[1] Wang, W., Song, Y., Zhang, J. & Deng, H., Automatic parking of vehicles: A review of literatures. *Int. J. Automot. Technol.*, **15**, pp. 967–978, 2014.

[2] Song, J., Zhang, W., Wu, X., Cao, H., Gao, Q. & Luo, S., Laser-based SLAM automatic parallel parking path planning and tracking for passenger vehicle. *IET Intell. Transp. Syst.*, **13**(10), pp. 1557–1568, 2019.

[3] Katrakazas, C., Quddus, M., Chen, W.-H. & Deka, L., Real-time motion planning methods for autonomous on-road driving: State-of-the-art and future research directions. *Transp. Res. Part C Emerg. Technol.*, **60**, pp. 416–442, 2015.

[4] Chen, Q., Gan, L., Chen, B., Liu, Q. & Zhang, X., Parallel parking path planning based on improved arctangent function optimization. *Int. J. Automot. Technol.*, **24**(1), pp. 23–33, 2023.

[5] Ren, B.T. et al., An automatic parking path optimization method based on hybrid A* and variable radius RS curve. *Chinese J. Highw.*, **35**(07), pp. 317–327, 2022.

[6] Zhang, J. et al., Autonomous valet parking path planning based on hierarchical architecture. *J. Southeast Univ. (Natural Sci. Ed.)*, **51**(5), pp. 8–887, 2021.

[7] Su, B., Yang, J., Li, L. & Wang, Y., Secondary parallel automatic parking of endpoint regionalization based on genetic algorithm. *Cluster Comput.*, **22**, pp. 7515–7523, 2019.

[8] Vorobieva, H., Glaser, S., Minoiu-Enache, N. & Mammar, S., Geometric path planning for automatic parallel parking in tiny spots. *IFAC Proc.* **45**(24), pp. 36–42, 2012.

[9] Chen, J., Lan, X. & Chen, F., Autonomous parking path planning based on improved deep reinforcement learning. *J. Chongqing Univ. Technol. (Natural Sci.)*, **35**(7), pp. 17–27, 2021.

[10] Shin, H., Kim, M.-J., Baek, S., Crane, C.D. & Kim, J., Perpendicular parking path generation and optimal path tracking algorithm for auto-parking of trailers. *Int. J. Control. Autom. Syst.*, **20**(9), pp. 3006–3018, 2022.

[11] Wu, Y., Li, X., Gao, J. & Yang, X., Research on automatic vertical parking path-planning algorithms for narrow parking spaces. *Electronics*, **12**(20), p. 4203, 2023.

[12] Wang, W. & Hou, Z., Model-free adaptive control based automatic parking scheme. *Control Decis.*, **37**(8), pp. 2056–2066, 2022.

[13] Horváth, E., Hajdu, C. & KHorös, P., Novel pure-pursuit trajectory following approaches and their practical applications. *2019 10th IEEE International Conference on Cognitive Infocommunications (CogInfoCom)*, pp. 597–602, 2019.

[14] Nakrani, N. & Joshi, M., An intelligent fuzzy based hybrid approach for parallel parking in dynamic environment. *Procedia Comput. Sci.*, **133**, pp. 82–91, 2018.

[15] Daxwanger, W.A. & Schmidt, G.K., Skill-based visual parking control using neural and fuzzy networks. *1995 IEEE International Conference on Systems, Man and Cybernetics. Intelligent Systems for the 21st Century*, pp. 1659–1664, 1995.

[16] Scicluna, J., Gatt, N., Casha, E., Grech, O. & Micallef, I., FPGA-based autonomous parking of a car-like robot using fuzzy logic control. *2012 19th IEEE Int. Conf. Electron. Circuits Syst. (ICECS 2012)*, pp. 229–232, 2012.

[17] Cheng, K., Zhang, Y. & Chen, H., Planning and control for a fully-automatic parallel parking assist system in narrow parking spaces. *2013 IEEE Intelligent Vehicles Symposium (IV)*, pp. 1440–1445, 2013.

[18] Vorobieva, H., Glaser, S., Minoiu-Enache, N. & Mammar, S., Automatic parallel parking with geometric continuous-curvature path planning. *2014 IEEE Intelligent Vehicles Symposium Proceedings*, pp. 465–471, 2014.

[19] Vorobieva, H., Glaser, S., Minoiu-Enache, N. & Mammar, S., Automatic parallel parking in tiny spots: Path planning and control. *IEEE Trans. Intell. Transp. Syst.*, **16**(1), pp. 396–410, 2014.

[20] Naderi Samani, N., Danesh, M. & Ghaisari, J., Parallel parking of a car-like mobile robot based on the P-domain path tracking controllers. *IET Control Theory and Appl.*, **10**(5), pp. 564–572, 2016.

[21] Ji, X., Wang, J., Zhao, Y., Liu, Y., Zang, L. & Li, B., Path planning and tracking for vehicle parallel parking based on preview BP neural network PID controller. Trans. Tianjin Univ., 21(3, pp. 199–208, 2015.

[22] Li, B., Wang, K. & Shao, Z., Time-optimal maneuver planning in automatic parallel parking using a simultaneous dynamic optimization approach. *IEEE Trans. Intell. Transp. Syst.*, **17**(11), pp. 3263–3274, 2016.

[23] Wu, T.-F., Tsai, P.-S., Hu, N.-T. & Chen, J.-Y., Research and implementation of auto parking system based on ultrasonic sensors. *2016 International Conference on Advanced Materials for Science and Engineering (ICAMSE)*, pp. 643–645, 2016.

[24] Nedamani, H.R., Khiabani, P.M. & Azadi, S., Intelligent parallel parking using adaptive neuro-fuzzy inference system based on fuzzy C-means clustering algorithm. *SAE Technical Papers*, 2018.

[25] Lee, B., Wei, Y. & Guo, I.Y., Automatic parking of self-driving car based on lidar. *Int. Arch. Photogramm. Remote Sens. Spat. Inf. Sci.*, **42**, pp. 241–246, 2017.

[26] Qin, Z., Chen, X., Hu, M., Chen, L. & Fan, J., A novel path planning methodology for automated valet parking based on directional graph search and geometry curve. *Rob. Auton. Syst.*, **132**, 103606, 2020.

[27] Zips, P., Böck, M. & Kugi, A., Optimisation based path planning for car parking in narrow environments. *Rob. Auton. Syst.*, **79**, pp. 1–11, 2016.

[28] Banzhaf, H., Sanzenbacher, P., Baumann, U. & Zöllner, J.M., Learning to predict ego-vehicle poses for sampling-based nonholonomic motion planning. *IEEE Robot. Autom. Lett.*, **4**(2), pp. 1053–1060, 2019.

[29] Chen, C., Rickert, M. & Knoll, A., Path planning with orientation-aware space exploration guided heuristic search for autonomous parking and maneuvering. *2015 IEEE Intelligent Vehicles Symposium (IV)*, pp. 1148–1153, 2015.

[30] Lee, S., Lim, W. & Sunwoo, M., Robust parking path planning with error-adaptive sampling under perception uncertainty. *Sensors*, **20**(12), p. 3560, 2020.

[31] Fan, Z. & Chen, H., Study on path following control method for automatic parking system based on LQR. *SAE Int. J. Passeng. Cars-Electronic Electr. Syst.*, **10**(2016-01–1881), pp. 41–49, 2016.

[32] Du, X. & Tan, K.K., Autonomous reverse parking system based on robust path generation and improved sliding mode control. *IEEE Trans. Intell. Transp. Syst.*, **16**(3), pp. 1225–1237, 2014.

[33] Ballinas, E., Montiel, O., Castillo, O., Rubio, Y. & Aguilar, L.T., Automatic parallel parking algorithm for a carlike robot using fuzzy PD+ I control. *Eng. Lett.*, **26**(4), 2018.

[34] Chen, C.-Y. & Feng, H.-M., Hybrid intelligent vision-based car-like vehicle backing systems design. *Expert Syst. Appl.*, **36**(4), pp. 7500–7509, 2009.

[35] Joshi, M.M. & Zaveri, M.A., Reactive navigation of autonomous mobile robot using neuro-fuzzy system. *Int. J. Robot. Autom.*, **2**(3), p. 128, 2011.
[36] Marasigan, A.A., Saberon, J.T., San Jose, I.M.B., Sevilla, D.P.B. & Bandala, P.A.T., Autonomous parallel parking of four wheeled vehicles utilizing adoptive fuzzy-neuro control system. *2014 IEEE Reg. 10 Symp.*, pp. 640–644, 2014.
[37] Aye, Y.Y., Watanabe, K., Maeyama, S. & Nagai, I., An intelligent parking system for vehicles using an image-based fuzzy controller. *Int. J. Smart Mater. Mechatronics*, **4**(1), 2017.
[38] Nari, M.I., Mustain, Z., Kautsar, S. & Utomo, S.B., Parallel parking system design with fuzzy logic control. *J. Nas. Tek. Elektro*, **10**(2), 2021. https://doi.org/10.25077/jnte.v10n2.904.2021.
[39] Heinen, M.R., Osório, F.S., Heinen, F.J. & Kelber, C., Autonomous vehicle parking and pull out using artificial neural networks. *Proceedings of the I Workshop on Computational Intelligence (WCI)*, 2006.
[40] Farooq, U., Amar, M., Asad, M.U., Hanif, A. & Saleh, S.O., Design and implementation of neural network based controller for mobile robot navigation in unknown environments. *Int. J. Comput. Electr. Eng.*, **6**(2), pp. 83–89, 2014.
[41] Zhou, W., An improved approach for automatic parallel parking in narrow parking spaces. 2015.
[42] Liu, W., Li, Z., Li, L. & Wang, F.-Y., Parking like a human: A direct trajectory planning solution. *IEEE Trans. Intell. Transp. Syst.*, **18**(12), pp. 3388–3397, 2017.
[43] Li, R., Wang, W., Chen, Y., Srinivasan, S. & Krovi, V.N., An end-to-end fully automatic bay parking approach for autonomous vehicles. *Dynamic Systems and Control Conference*, V002T15A004, 2018.
[44] Moon, J., Bae, I. & Kim, S., Automatic parking controller with a twin artificial neural network architecture. *Math. Probl. Eng.*, **2019**(1), 4801985, 2019.
[45] Ma, Y. & Wang, Z., Artificial intelligence applications in the development of autonomous vehicles: A survey, 2020.
[46] Ghaseminezhad, M.H. & Karami, A., A novel self-organizing map (SOM) neural network for discrete groups of data clustering. *Appl. Soft Comput.*, **11**(4), pp. 3771–3778, 2011.
[47] De Luca, A., Oriolo, G. & Samson, C., Feedback control of a nonholonomic car-like robot. *Robot Motion Plan. Control*, pp. 171–253, 2005.

# LOCAL MOVING LEAST SQUARES METHOD FOR GRANULAR DEBRIS FLOW

FILIP CIGÁŇ & JURAJ MUŽÍK
University of Žilina, Slovak Republic

## ABSTRACT

This study investigates the potential applications of local meshless methods in modelling rapid slope movements dynamics. These geohazards pose significant threats to mountainous infrastructure, necessitating the development of robust numerical models for predictive purposes. Contemporary research has seen an increased focus on meshless methods across various disciplines due to their potential advantages over traditional numerical approaches. In this investigation, the local weighted residual method was employed to develop a debris flow model, as it demonstrated suitability for addressing the relevant differential equations. To assess the efficacy of this methodology, the research centred on dry granular soil movement models, with the resultant solutions being compared against several established case studies. This comparative analysis aims to validate the viability of the proposed method in the context of geohazard modelling.

*Keywords: debris flow, meshless method, moving least squares.*

## 1 INTRODUCTION

Debris flows, avalanches and rapid mass movements represent significant geohazards in mountainous regions worldwide. These phenomena, characterised by the rapid downslope movement of granular materials mixed with varying amounts of water, pose substantial threats to infrastructure and human lives. The complexity of these flows, arising from their nonlinear dynamics, variable rheology, and interaction with complex topographies, necessitates advanced numerical modelling techniques for accurate prediction and risk assessment. The foundation for many contemporary debris flow models lies in the seminal work of Savage and Hutter [1], who proposed a theoretical framework for describing the rapid creep of granular materials along inclined surfaces in 1989. This theory and its subsequent extensions [2], [3] form the basis of depth-averaged mathematical models that capture the essential physics of debris flows while remaining computationally tractable. The governing equations for debris flows are similar to the shallow water equations (SWEs) but with additional terms for internal friction and variable basal topography. These equations form a system of hyperbolic partial differential equations (PDEs) that can exhibit shock-like phenomena and require careful numerical treatment. Traditionally, finite volume methods [4] have been employed to solve these equations. These methods offer several advantages, including ease of implementation for complex geometries, the natural handling of large deformations, and the ability to adapt to resolution dynamically. Those based on radial basis functions (RBFs) have gained popularity among meshless [5] techniques. However, RBF methods often lead to ill-conditioned matrices, particularly for large-scale problems. In this study, we propose and investigate the application of the local moving least squares (LMLS) method [6] to debris flow modelling. The LMLS approach offers several key advantages.

It provides a robust framework for spatial discretisation without a structured mesh. It allows for high-order approximations, which are crucial for capturing the complex dynamics of debris flows. It maintains good conditioning even for large numbers of computational points. It naturally handles irregular point distributions, enabling adaptive resolution in areas of interest.

WIT Transactions on Engineering Sciences, Vol 136, © 2024 WIT Press
www.witpress.com, ISSN 1743-3533 (on-line)
doi:10.2495/BE470061

Our numerical scheme combines the LMLS spatial discretisation with a high-order Runge–Kutta time integration method. This approach is designed to capture the debris flow nonlinear behaviour accurately while maintaining numerical stability and efficiency. The remainder of this paper is structured as follows.

Section 2 presents the governing equations based on the extended Savage–Hutter theory, detailing the depth-averaged mass and momentum conservation laws with Mohr–Coulomb rheology. Section 3 describes our numerical methodology, including the temporal discretisation scheme (Section 3.1), the LMLS spatial approximation technique (Section 3.2) and a solution stabilisation process to enhance robustness (Section 3.3).

Section 4 demonstrates the efficacy of our approach through a comprehensive numerical example, simulating the flow of dry cohesionless soil down an inclined slope. Section 5 concludes the paper with a summary of our findings and directions for future research.

Through this work, we aim to contribute to the ongoing development of accurate and efficient numerical tools for debris flow modelling, ultimately enhancing our ability to assess and mitigate related geohazards in mountainous regions.

## 2 GOVERNING EQUATIONS

This study's mathematical model for debris flows is based on the extended Savage–Hutter theory, which combines principles from SWEs with a Mohr–Coulomb rheology for granular materials. This approach simulates rapid, gravity-driven flows of cohesionless granular materials over complex topographies.

The soil body can be modelled at yield as a Mohr–Coulomb plastic material that slides over a rigid basal topography. The depth integration decreases the solution's dimension to two-dimensional planar movement.

### 2.1 Definition of the curvilinear coordinate system

The curvilinear coordinate system expresses the governing equations (Fig. 1). The depth-integrated mass balance equation can be written as:

$$\frac{\partial h}{\partial t} + \frac{\partial (hu)}{\partial x} + \frac{\partial (hv)}{\partial y} = 0 \tag{1}$$

and the moment balance equations are:

$$\frac{\partial (hu)}{\partial t} + \frac{\partial (hu^2)}{\partial x} + \frac{\partial (huv)}{\partial y} = h\left(s_x - \beta_x \frac{\partial h}{\partial x}\right) \tag{2}$$

$$\frac{\partial (hv)}{\partial t} + \frac{\partial (huv)}{\partial x} + \frac{\partial (hv^2)}{\partial y} = h\left(s_y - \beta_y \frac{\partial h}{\partial y}\right) \tag{3}$$

where $h$ is the thickness of the avalanche, and $u$ and $v$ are the depth-averaged velocities in the directions $x$ and $y$, respectively. The terms $s_x$ and $s_y$ represent the accelerations in the downslope and lateral directions. If the basal surface is curved downslope and laterally planar, these terms are defined as [4]:

$$s_x = g\left(\sin\xi - \cos\xi \frac{u}{|\mathbf{u}|}\tan\delta\right) \tag{4}$$

$$s_y = -g\cos\xi \frac{v}{|\mathbf{u}|}\tan\delta \tag{5}$$

$$\mathbf{u} = [u, v]^T, \quad |\mathbf{u}| = \sqrt{u^2 + v^2} \tag{6}$$

where $\delta$ is the basal Coulomb dry-friction angle.

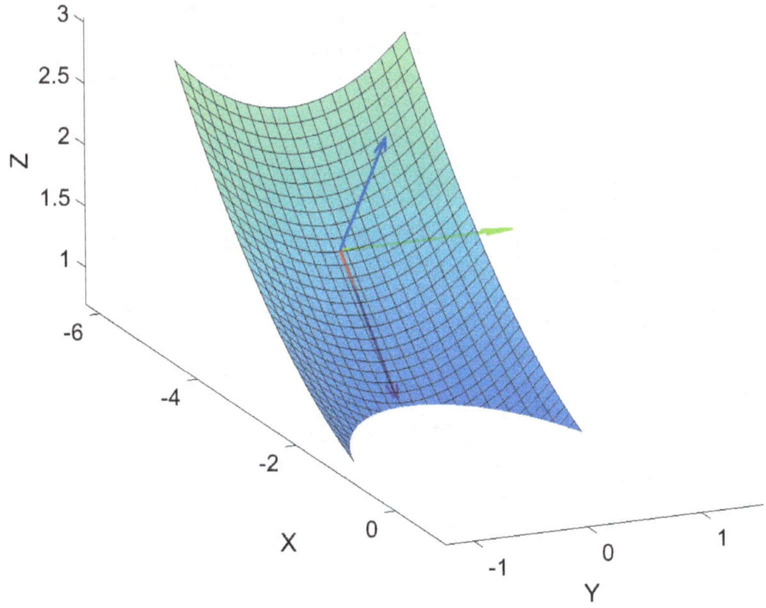


Figure 1: Definition of the curvilinear coordinate system.

The factors $\beta_x$ and $\beta_y$ are defined as [3]:

$$\beta_x = g\cos\xi K_x, \quad \beta_y = g\cos\xi K_y \tag{7}$$

where $\xi$ is the inclination and $K_x$, and $K_y$ are the down- and cross-slope earth pressure coefficients defined by the Mohr–Coulomb criterion:

$$K_{xa/xp} = \frac{2}{\cos^2\phi}\left(1 \mp \sqrt{1 - \sin^2\phi/\cos^2\delta}\right) - 1 \tag{8}$$

$$K_{ya/yp}^{xa/xp} = \frac{1}{2}\left(K_{xa/xp} + 1 \mp \sqrt{\left(K_{xa/xp} - 1\right)^2 + 4\tan^2\delta}\right) \tag{9}$$

where $\phi$ is the internal friction angle, and the subscripts '$a$' or '$p$' denote active or passive stress states [7]. The active stress states are generated in the case of soil dilation, and the passive stress states are joined with compression. The choice of active or passive state coefficients in a numerical model depends on the acceleration in an appropriate direction [4]:

$$K_x = \begin{cases} K_{xa} & \text{if } \partial u/\partial x \geq 0 \\ K_{xp} & \text{if } \partial u/\partial x < 0 \end{cases} \tag{10}$$

$$K_y = \begin{cases} K_{ya}^{xa} & if\partial u/\partial x \geq 0, \partial v/\partial y \geq 0 \\ K_{yp}^{xa} & if\partial u/\partial x \geq 0, \partial v/\partial y < 0 \\ K_{ya}^{xp} & if\partial u/\partial x < 0, \partial v/\partial y \geq 0 \\ K_{yp}^{xp} & if\partial u/\partial x < 0, \partial v/\partial y < 0 \end{cases} \tag{11}$$

The usual boundary conditions of eqns (1), (2) and (3) are:

- Given depth, where the value of the depth $h$ on the part of boundary $\Gamma_1$ is given, i.e., $h = h_0$,
- No-flux condition, where the flux perpendicular to the boundary $\Gamma_2$ is zero $q_n = 0$,
- No-slip condition, where the velocities $u$ and $v$ are zero on this part $\Gamma_3$ of the boundary.

and the whole boundary $\Gamma = \Gamma_1 \cup \Gamma_2 \cup \Gamma_3$. The initial condition is defined by the prescribed values of the depth and velocities at time $t_0 = 0$.

## 3 NUMERICAL SOLUTION METHODOLOGY

### 3.1 Temporal discretisation scheme

The governing equations for debris flow, derived from the extended Savage–Hutter theory, form a system of hyperbolic PDEs. In vector form, these can be expressed as:

$$\frac{\partial \mathbf{U}}{\partial t} + \frac{\partial \mathbf{F}(\mathbf{U})}{\partial x} + \frac{\partial \mathbf{G}(\mathbf{U})}{\partial y} = \mathbf{S}(\mathbf{U}) \tag{12}$$

where $\mathbf{U}$ is the vector of conserved variables, $\mathbf{F}(\mathbf{U})$ and $\mathbf{G}(\mathbf{U})$ are the flux tensors, and $\mathbf{S}(\mathbf{U})$ represents the source terms. Specifically:

$$\mathbf{U} = \begin{bmatrix} h \\ hu \\ hv \end{bmatrix}, \quad \mathbf{F}(\mathbf{U}) = \begin{bmatrix} hu \\ hu^2 \\ huv \end{bmatrix}, \quad \mathbf{G}(\mathbf{U}) = \begin{bmatrix} hv \\ huv \\ hv^2 \end{bmatrix}, \quad \mathbf{S}(\mathbf{U}) = \begin{bmatrix} 0 \\ h\left(s_x - \beta_x \frac{\partial h}{\partial x}\right) \\ h\left(s_y - \beta_y \frac{\partial h}{\partial y}\right) \end{bmatrix} \tag{13}$$

Here, $h$ is the flow depth, $u_x$ and $u_y$ are depth-averaged velocities in the $x$ and $y$ directions, respectively, $g$ is the gravitational acceleration, and $\kappa_x$ and $\kappa_y$ are earth pressure coefficients.

To solve this system numerically, we employ an explicit third-order Runge–Kutta (RK3) method for time integration. This choice balances computational efficiency with the accuracy required for capturing the complex dynamics of debris flows. The RK3 scheme is formulated as follows:

$$\begin{aligned} \mathbf{U}^{(0)} &= \mathbf{U}^n \\ \mathbf{U}^{(1)} &= \mathbf{U}^{(0)} + \Delta t \mathbf{L}\left(\mathbf{U}^{(0)}\right) \\ \mathbf{U}^{(2)} &= \frac{3}{4}\mathbf{U}^{(0)} + \frac{1}{4}\left[\mathbf{U}^{(1)} + \Delta t \mathbf{L}\left(\mathbf{U}^{(1)}\right)\right] \\ \mathbf{U}^{(3)} &= \frac{1}{3}\mathbf{U}^{(0)} + \frac{2}{3}\left[\mathbf{U}^{(2)} + \Delta t \mathbf{L}\left(\mathbf{U}^{(2)}\right)\right] \\ \mathbf{U}^{n+1} &= \mathbf{U}^{(3)} \end{aligned} \tag{14}$$

where $\Delta t$ is the time step, and the operator $\mathbf{L}(\mathbf{U})$ represents the spatial discretisation. The spatial derivatives in $\mathbf{L}(\mathbf{U})$ are approximated using the LMLS method, as detailed in Section 3.2. This meshless approach allows for flexible handling of complex topographies often encountered in debris flow scenarios.

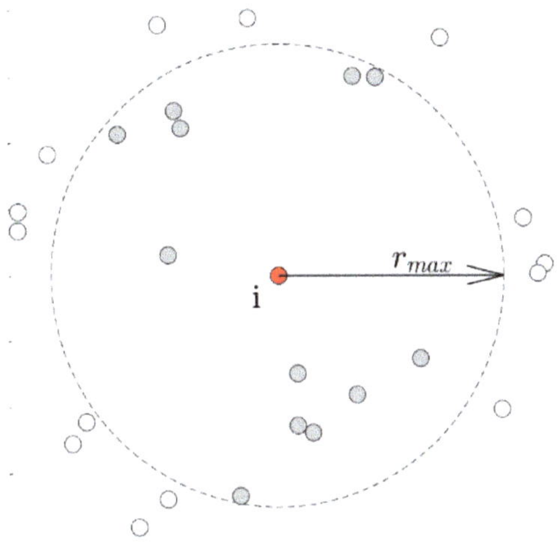


Figure 2: Support domain of the *i*th point.

### 3.2 Spatial derivative approximation

We employ the LMLS method to calculate spatial derivatives in our debris flow model. This meshless approach offers enhanced accuracy and stability compared to traditional methods, particularly in preserving mass conservation.

In the LMLS method, we approximate the function $\phi(x,y)$ at any point $(x,y)$ within the domain as:

$$\phi(x,y) \approx \sum_{j=1}^{m} \alpha_j\,(x,y)p_j(x,y) \quad (15)$$

where $p_j(x,y)$ are polynomial basis functions, and $\alpha_j(x,y)$ are unknown coefficients that depend on the spatial coordinates. The choice of $m$ corresponds to the polynomial degree, with the 2D polynomial set typically defined as:

$$\{p_j(x,y)\} = \{1, x, y, x^2, xy, y^2, \ldots\} \quad (16)$$

To determine the unknown coefficients $\alpha_j(x,y)$, we minimise the following weighted least squares functional:

$$J(\mathbf{x}) = \sum_{i=1}^{n} W\,(\mathbf{x}-\mathbf{x}_i)\left[\phi_i - \sum_{j=1}^{m} \alpha_j\,(\mathbf{x})p_j(\mathbf{x}_i)\right]^2 \quad (17)$$

where $\mathbf{x} = (x,y)$, $\mathbf{x}_i = (x_i,y_i)$ are the scattered data points, $\phi_i$ are the known function values at these points, and $W(\mathbf{x}-\mathbf{x}_i)$ is a weight function.

A common choice for the weight function is the exponential function:

$$W(\mathbf{x}-\mathbf{x}_i) = \exp(-\beta \parallel \mathbf{x}-\mathbf{x}_i \parallel^2) \quad (18)$$

where $\beta$ is a shape parameter controlling the influence of neighbouring points.

Minimising the functional $J(\mathbf{x})$ with respect to $\alpha_j(\mathbf{x})$ leads to the following system of equations:

$$\sum_{j=1}^{m} \alpha_j\,(\mathbf{x}) \sum_{i=1}^{n} W\,(\mathbf{x}-\mathbf{x}_i)p_j(\mathbf{x}_i)p_k(\mathbf{x}_i) = \sum_{i=1}^{n} W\,(\mathbf{x}-\mathbf{x}_i)\phi_i p_k(\mathbf{x}_i) \quad (19)$$

for $k = 1, \ldots, m$. This can be written in matrix form as:

$$\mathbf{A}(\mathbf{x})\boldsymbol{\alpha}(\mathbf{x}) = \mathbf{B}(\mathbf{x})\boldsymbol{\phi} \tag{20}$$

where:

$$\begin{aligned} A_{kj}(\mathbf{x}) &= \textstyle\sum_{i=1}^{n} W(\mathbf{x} - \mathbf{x}_i) p_j(\mathbf{x}_i) p_k(\mathbf{x}_i) \\ B_k(\mathbf{x}) &= \textstyle\sum_{i=1}^{n} W(\mathbf{x} - \mathbf{x}_i) p_k(\mathbf{x}_i) \end{aligned} \tag{21}$$

The solution for the unknown coefficients is then:

$$\boldsymbol{\alpha}(\mathbf{x}) = \mathbf{A}^{-1}(\mathbf{x})\mathbf{B}(\mathbf{x})\boldsymbol{\phi} \tag{22}$$

Once we have determined $\boldsymbol{\alpha}(\mathbf{x})$, we can approximate the function and its derivatives at any point $\mathbf{x}$ as:

$$\begin{aligned} \phi(\mathbf{x}) &\approx \textstyle\sum_{j=1}^{m} \alpha_j(\mathbf{x}) p_j(\mathbf{x}) \\ \frac{\partial \phi}{\partial x}(\mathbf{x}) &\approx \textstyle\sum_{j=1}^{m} \alpha_j(\mathbf{x}) \frac{\partial p_j}{\partial x}(\mathbf{x}) \\ \frac{\partial \phi}{\partial y}(\mathbf{x}) &\approx \textstyle\sum_{j=1}^{m} \alpha_j(\mathbf{x}) \frac{\partial p_j}{\partial y}(\mathbf{x}) \end{aligned} \tag{23}$$

### 3.3 Solution stabilisation process

While the LMLS method described in Section 3.2 provides a robust framework for spatial discretisation, the time integration process may still introduce numerical instabilities, particularly in regions with steep gradients or discontinuities standard in debris flows. To address this, we implement a solution stabilisation process that leverages the inherent properties of the LMLS approximation.

Our stabilisation approach focuses on the depth variable $h$, critical for mass conservation in debris flow simulations. At each time step $n + 1$, we have two values for depth at each point $i$:

1. $h_i^{n+1}$: obtained directly from the Runge–Kutta time integration (as described in Section 3.1).
2. $h'^{n+1}_i$: the LMLS approximation computed using the coefficients $\alpha_j(\mathbf{x}_i)$ from Section 3.2.

We define a local error measure $\varepsilon_i$ at each point $i$:

$$\varepsilon_i = \frac{|h_i^{n+1} - h'^{n+1}_i|}{h'^{n+1}_i} \tag{24}$$

We compare this local error to a threshold value $\varepsilon_{threshold}$ to determine whether to apply smoothing. We use the following adaptive smoothing algorithm:

1. If $\varepsilon_i > \varepsilon_{threshold}$:

$$h_i^{n+1} = \omega h_i^{n+1} + (1 - \omega) h'^{n+1}_i \tag{25}$$

2. If $\varepsilon_i \leq \varepsilon_{threshold}$: No smoothing is applied.

Here, $\omega$ is a relaxation parameter ($0 < \omega < 1$) that controls the degree of smoothing. This stabilisation process takes advantage of the LMLS approximation's continuity and smoothness properties while preserving sharp features when the local error is below the threshold. It helps maintain numerical stability without overly diffusing the solution, which is crucial for accurately capturing the complex dynamics of debris flows.

The adaptive nature of this stabilisation technique is particularly beneficial in handling the varying flow regimes encountered in debris flow simulations, from rapid, turbulent flows to slower, more viscous behaviours. By applying smoothing selectively, we can maintain the sharp flow fronts and discontinuities characteristic of debris flows while dampening numerical oscillations in smoother flow regions.

It's important to note that this stabilisation process is applied after each Runge–Kutta substep, ensuring stability is maintained throughout the time-integration process. The LMLS coefficients $\alpha_j(\mathbf{x}_i)$ are recomputed at each time step using the updated depth values, maintaining consistency between the stabilised solution and the LMLS approximation.

This approach balances numerical stability and solution accuracy, crucial for reliable long-term simulations of complex geophysical flows. It complements the spatial discretisation provided by the LMLS method, resulting in a robust numerical scheme capable of handling the challenges posed by debris flow modelling.

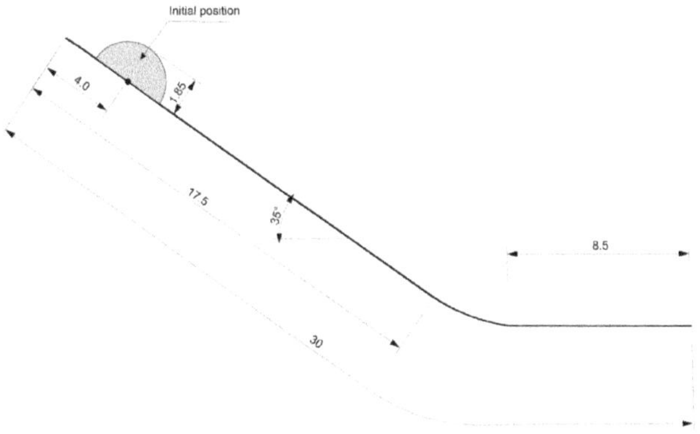


Figure 3: Longitudinal profile of computational area at $y = 0$ with initial position of the soil material [5].

## 4 NUMERICAL CASE STUDY: DRY COHESIONLESS SOIL FLOW ON AN INCLINED SLOPE

This section presents a comprehensive numerical example that demonstrates the capabilities and efficacy of the LMLS method in simulating a complex debris flow scenario. The example focuses on modelling the movement of a given volume of dry cohesionless soil down an inclined slope [5].

### 4.1 Problem setup

The computational domain consists of a planar slope inclined at an angle $\xi(x)$ that varies with the x-coordinate according to:

$$\xi(x) = \begin{cases} 35^\circ & \text{for } 0 \le x \le 17.5 \\ 35^\circ\left(1 - \left(\frac{x-17.5}{4}\right)^2\right) & \text{for } 17.5 < x < 21.5 \\ 0^\circ & \text{for } x \ge 21.5 \end{cases} \quad (26)$$

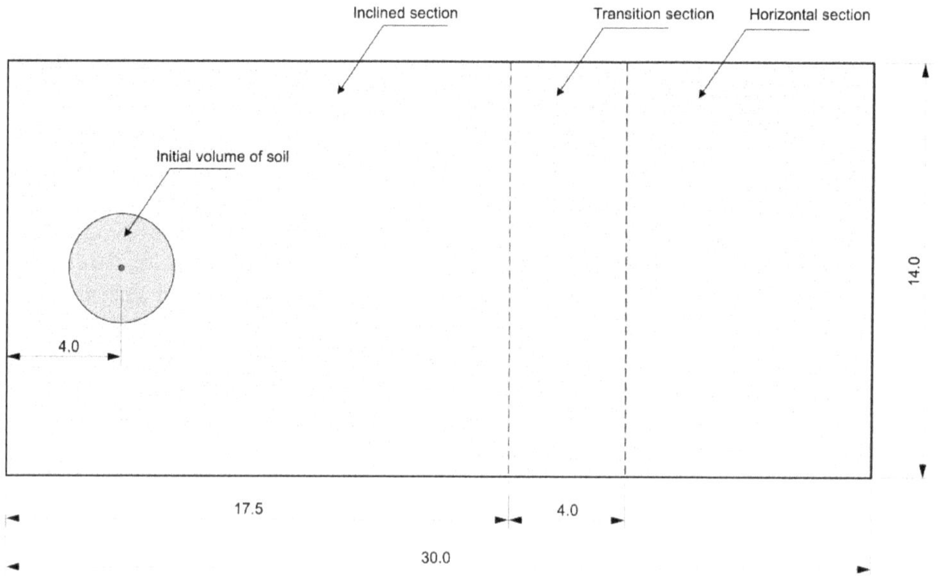


Figure 4: Dimensions of computational area, network of points and initial position of the soil material [5].

This configuration represents a steep upper slope transitioning smoothly to a horizontal runout zone, mimicking typical topographies encountered in debris flow hazard areas.

The initial soil mass is modelled as a hemisphere with radius $R = 1.85$, centred at coordinates $(x_0, y_0) = (4{,}0)$. The material properties are defined as follows:

- Internal friction angle: $\phi = 30°$.
- Bed friction angle: $\delta = 30°$.

### 4.2 Numerical implementation

We employ two distinct point distributions to discretise the computational domain [5]:

1. A regular rectangular grid comprising 10,721 points (151 × 71).
2. An irregular distribution containing 15,700 points, generated using the Poisson disc algorithm [5].

In the LMLS approximation, we use quadratic basis functions ($m = 6$) for both distributions. The support size for each point is adaptively determined to include approximately 50 neighbouring points. The simulation is run for a dimensionless time of $T = 24$.

### 4.3 Results and discussion

Figs 5 and 6. illustrates the evolution of the flow depth contours at dimensionless times $t = 3, 6, 9, 12, 15, 18, 21$ and 24 for both the regular and irregular point distributions.

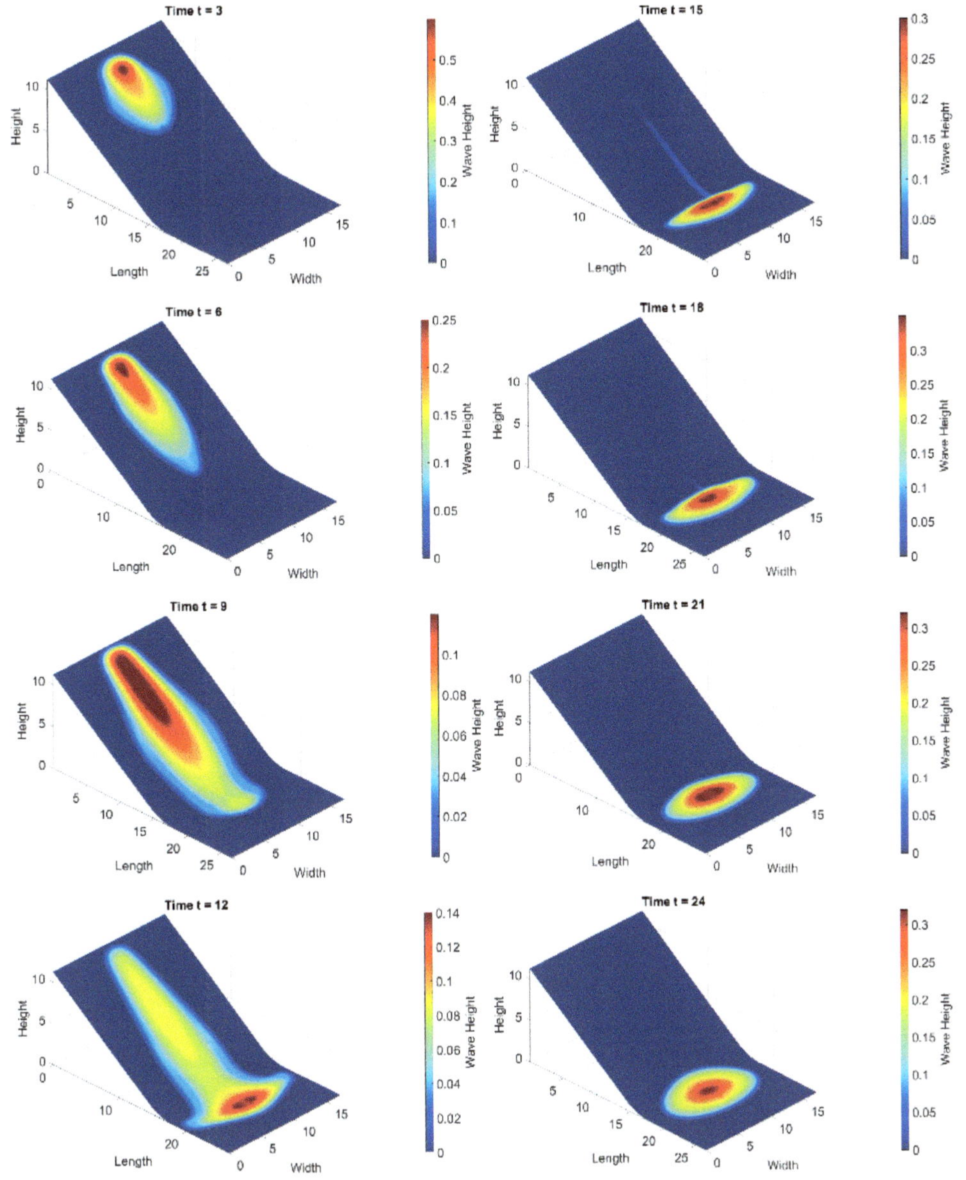


Figure 5: Regular grid, contours of depth.

Key observations include:

1. Rapid deformation and elongation of the initial hemispherical mass as it accelerates down the steep upper slope.
2. Development of a pronounced flow front, characterised by a sharp increase in flow depth.

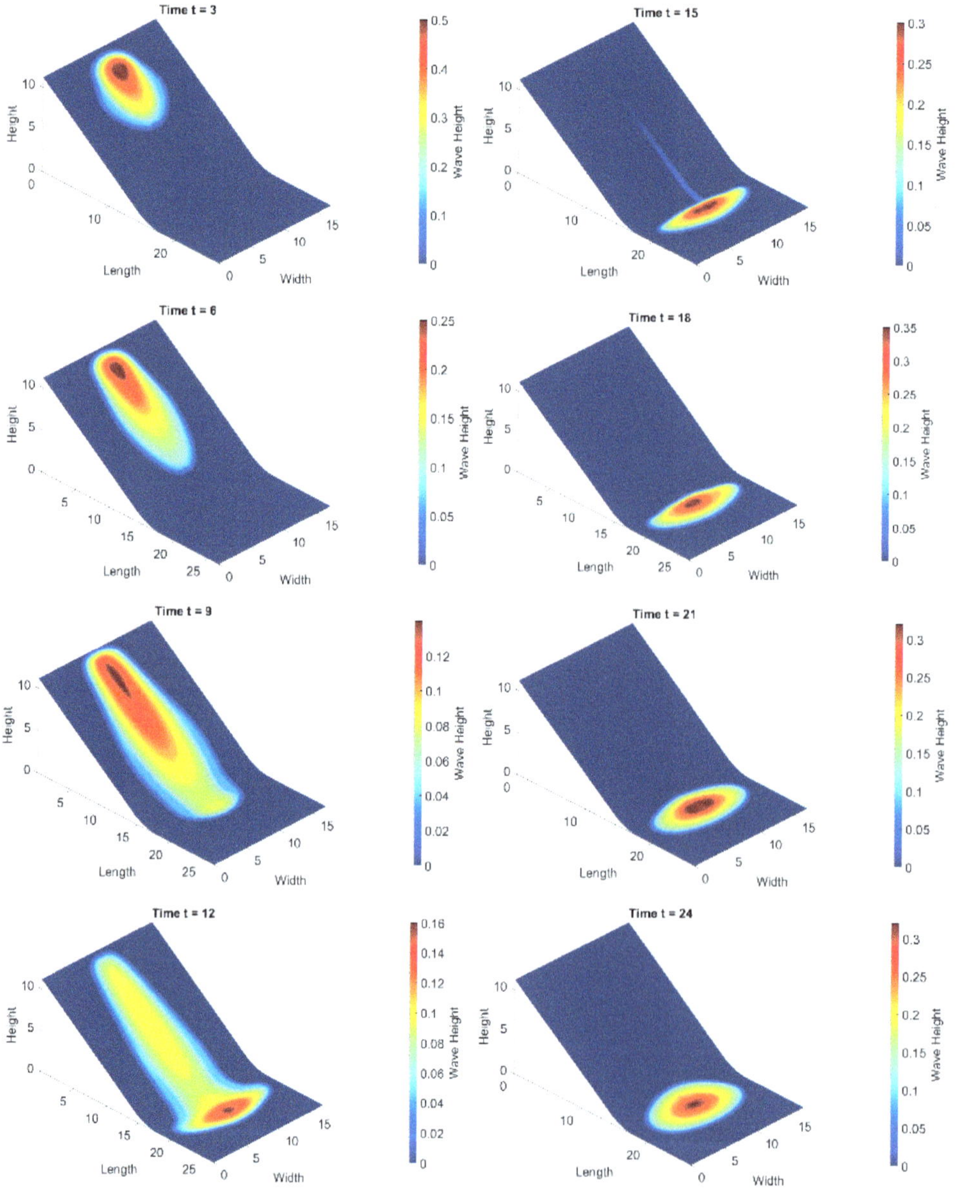


Figure 6: Irregular grid, contours of depth.

3. Deceleration and lateral spreading as the material enters the transition zone.
4. Continued spreading and gradual cessation of motion on the horizontal runout zone.

To assess the conservation properties of our numerical scheme (Fig. 7), we compute the relative error in total material volume over time:

$$\epsilon_V(t) = \frac{|V(t)-V(0)|}{V(0)} \tag{27}$$

where $V(t)$ is the total volume at time $t$.

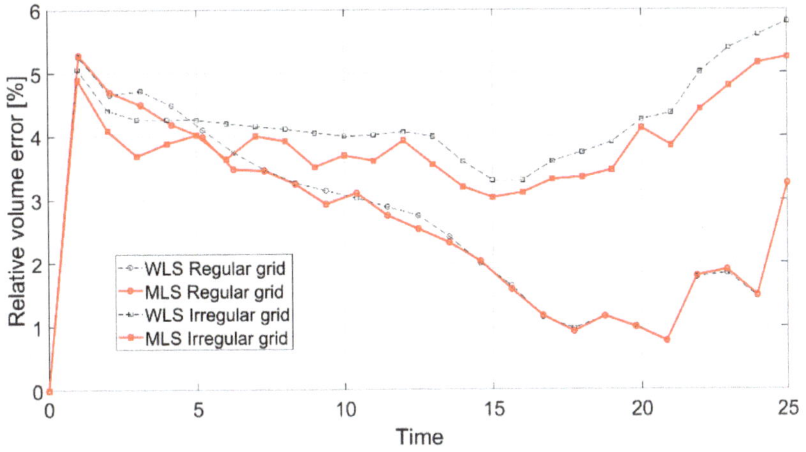


Figure 7: Course of the relative volume error over time. Comparison of MLS and WLS for regular and irregular grid.

## 5 CONCLUSIONS AND FUTURE WORK

This study has presented a robust numerical framework for modelling debris flows using the LMLS method. Our approach combines the extended Savage–Hutter theory for granular flows with a meshless spatial discretisation technique, offering several key advantages in simulating these complex geophysical phenomena.

The LMLS method, coupled with an RK3 time integration scheme, demonstrates excellent capability in capturing the nonlinear dynamics of debris flows. The numerical case study of dry cohesionless soil flow on an inclined slope showcases the method's ability to accurately model flow initiation, propagation, and deposition phases. Our approach exhibits strong conservation properties. The LMLS method's flexibility in handling regular and irregular point distributions is a significant advantage. This feature allows for adaptive resolution in areas of complex topography or high-flow gradients without complex remeshing algorithms. The proposed solution stabilisation process effectively mitigates numerical oscillations while preserving sharp flow features, which is essential for accurately representing the shock-like phenomena often observed in debris flows. The method's computational efficiency is demonstrated by its ability to simulate complex flows over extended periods on standard desktop hardware, making it accessible for research and practical engineering applications.

## ACKNOWLEDGEMENTS

This contribution is the result of the project funded by the Scientific Grant Agency of Slovak Republic (VEGA) No. 1/0752/24 'Research of composite foamed concrete-based panels for applications in traffic construction' and 1/0009/23 'Numerical and experimental modelling of dynamic phenomena on the constructions of transport structures and their surroundings'.

## REFERENCES

[1] Savage, S.B. & Hutter, K., The motion of a finite mass of granular material down a rough incline. *Journal of Fluid Mechanics*, **199**, pp. 37–68, 1989.
[2] Hutter, K., Siegel, M., Savage, S.B. & Nohguchi, Y., Two-dimensional spreading of a granular avalanche down an inclined plane: Part I – Theory. *Acta Mechanica*, **100**, pp. 37–68, 1993.
[3] Pudasaini, S.P. & Hutter, K., Rapid shear flows of dry granular masses down curved and twisted channels. *Journal of Fluid Mechanics*, **495**, pp. 193–208, 2003.
[4] Wang, Y., Hutter, K. & Pudasaini, S.P., The Savage–Hutter theory: A system of partial differential equations for avalanche flows of snow, debris, and mud. *ZAMM Zeitschrift fur Angewandte Mathematik und Mechanik*, **84**, pp. 507–527, 2004.
[5] Kovářík, K., Mužík, J., Masarovičová, S., Bulko, R. & Gago, F., The local meshless numerical model for granular debris flow. *Engineering Analysis with Boundary Elements*, **130**, pp. 20–28, 2021.
[6] Liu, G.R. & Gu, Y.T., *An Introduction to Meshfree Methods and Their Programming*, Springer.
[7] Hsu, T.-W., Liang, S.-J. & Wu, N.-J., Application of meshless SWE model to moving wet/dry front problems. *Engineering with Computers*, **35**, pp. 291–303, 2019.

# A MESHLESS COLLOCATION METHOD BASED ON HIERARCHICAL MATRICES FOR EFFICIENT NUMERICAL ANALYSIS OF LARGE-SCALE LINEAR SYSTEMS

MINGJUAN LI, WENZHI XU & ZHUOJIA FU
College of Mechanics and Engineering Sciences, Hohai University, China

## ABSTRACT

This paper presents a novel approach to solving large-scale linear algebra problems by integrating hierarchical matrices (H-matrices) with a semi-analytic meshless collocation method (singular boundary method). The singular boundary method eliminates the need for mesh generation by using a set of discrete nodes to approximate the solution of partial differential equations. However, this approach often leads to dense linear systems that are computationally expensive and memory-intensive. To address this challenge, we employ H-matrices, which decompose these dense matrices into hierarchical sub-blocks that can often be approximated by low-rank matrices. This decomposition significantly reduces computational complexity from the conventional to or lower, depending on the matrix structure. The implementation of these data-sparse representations significantly enhances the efficiency of numerical computations by reducing both complexity and storage requirements traditionally associated with matrix operations. Our results demonstrate that the combination of H-matrices with the singular boundary method not only reduces memory requirements but also accelerates matrix operations, such as matrix-vector multiplication and inversion. The validity of the proposed method for solving the Laplace equation and the Helmholtz equation is verified by benchmark examples, highlighting its potential for large-scale engineering problems.

*Keywords: meshless collocation method, singular boundary method, hierarchical matrices, large-scale linear algebra, low-rank approximation.*

## 1 INTRODUCTION

Over the past few decades, various numerical methods have been developed to solve partial differential equations (PDEs) [1], [2], thereby enabling a deeper understanding of their physical and mathematical properties. Among these, semi-analytic meshless collocation methods have been widely employed to solve a range of complex problems. The singular boundary method (SBM) [3]–[5], which is derived from the boundary element method [6], [7] and the method of fundamental solutions (MFS) [8], is a boundary-type meshless collocation method. SBM introduces the concept of source intensity factors, effectively overcoming the fictitious boundary issues inherent in the MFS while retaining key advantages of boundary-type numerical methods, such as the elimination of numerical integration and ease of implementation. These characteristics have led to the widespread application of SBM across various fields.

Despite the numerous advantages of the SBM, the method typically produces dense and non-symmetric coefficient matrices, resulting in significant computational costs and high memory demands. This makes the efficient solution of large-scale problems using SBM a substantial challenge. To address these issues, this paper proposes a novel approach to solving large-scale linear algebra problems by integrating hierarchical matrices (H-matrices) [9]–[11] with a semi-analytic meshless collocation method (singular boundary method). H-matrices represent a state-of-the-art technique designed to efficiently handle large, dense matrices that arise during the discretisation of PDEs. They offer a data-sparse representation


www.witpress.com, ISSN 1743-3533 (on-line)
doi:10.2495/BE470071

that is particularly effective in capturing non-local interactions within complex problems. Traditional matrix representations are computationally expensive and memory-intensive, but H-matrices differ by approximating distant matrix blocks with low-rank matrices, thus reducing both computational complexity and storage requirements. The core idea behind H-matrices is to partition a potentially sparse large matrix into smaller sub-blocks, which can then be approximated by low-rank matrices. This hierarchical representation significantly reduces the computational cost associated with large-scale matrix operations while maintaining the desired level of accuracy.

The integration of H-matrices with SBM effectively addresses the challenges posed by dense matrices in large-scale systems. As the scale of the problem increases, the computational burden associated with solving these systems also rises. By reducing the data complexity involved in matrix storage and operations, H-matrices enable SBM to solve more complex problems efficiently. The computational results presented in this paper demonstrate that this combined approach offers a promising solution for the efficient numerical analysis of large-scale linear systems, with broad applicability to problems such as the Laplace and Helmholtz equations.

## 2 METHODOLOGY

The SBM is a meshless collocation technique that eliminates the need for traditional mesh generation by using discrete collocation nodes to approximate the solution of PDEs. However, SBM often results in dense, non-symmetric matrices, which present significant challenges in terms of computational cost and memory requirements. To address these challenges, H-matrices are integrated with SBM, offering a data-sparse representation that mitigates the computational and storage demands associated with these dense systems.

### 2.1 Formulation of the SBM

To illustrate the methodology, consider the 3D Helmholtz equation in a bounded, homogeneous, isotropic domain $\Omega$:

$$(\Delta + \lambda^2)u = 0, \quad \text{in } \Omega, \tag{1}$$

subject to the boundary conditions:

$$u(x) = u_0(x), \quad x \in \Gamma_D, \tag{2}$$

$$\frac{\partial u(x)}{\partial n} = q(x), \quad x \in \Gamma_N, \tag{3}$$

where $u$ is the unknown function, $x$ represents the spatial coordinates, $\Omega$ and $\Gamma$ represent the computational domain and its boundary, with $\Gamma_D$ and $\Gamma_N$ representing Dirichlet and Neumann boundary conditions, respectively.

The SBM leads to the following linear system:

$$A_{ij}\alpha = b, \; A \in R^{n\times n}, x, b \in R^n, \tag{4}$$

where $A$ is a dense matrix,

$$A_{ij} = \begin{cases} \frac{exp(\sqrt{-1}\lambda r_{ij})}{4\pi r_{ij}}, & \text{if } i \neq j \\ \text{OIFs}, & \text{if } i = j \end{cases}, \tag{5}$$

with $r_{ij} = \|x_i - y_j\|$. The right-hand side vector $b$ corresponds to the boundary condition $u_0$.

When the collocation nodes $x_i$ coincide with the source nodes $y_j$, singularities arise in the fundamental solutions. SBM addresses these singularities by introducing origin intensity factors (OIFs) [12] to regularise the solution. OIFs can be determined by inverse interpolation techniques [13], subtraction and addition back techniques [14], [15] and integral mean values [16].

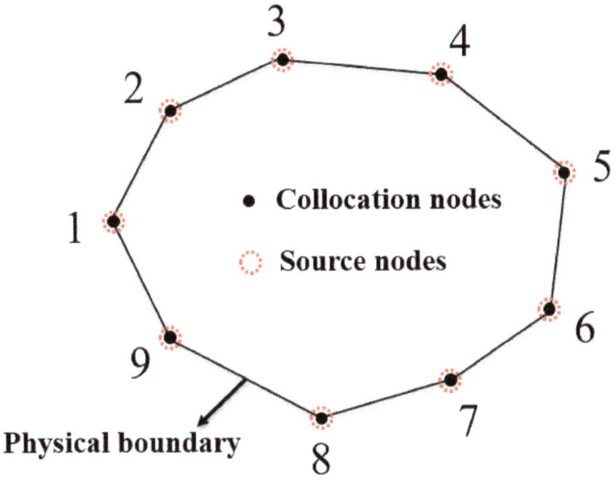


Figure 1: Schematic of the singular boundary method.

## 2.2 H-matrices for collocation method (SBM)

In the dense asymmetric matrix $A$ generated by SBM, directly solving the dense matrix $A \in R^{n\times n}$ requires $O(N^2)$ in storage and $O(N^3)$ in computational complexity. However, by leveraging H-matrices, the matrix can be divided into chunks where far-field interactions are represented by low-rank approximations.

Assuming the matrix $A$ has dimension $N$ and it consists of the following structure:

$$A = \begin{pmatrix} A_{11} & A_{12} & \cdots & A_{1N} \\ A_{21} & A_{22} & \cdots & A_{2N} \\ \vdots & \vdots & \ddots & \vdots \\ A_{N1} & A_{N2} & \cdots & A_{NN} \end{pmatrix}, \tag{6}$$

where certain sub-blocks (e.g., $A_{1N}$ $A_{N1}$ $A_{2N}$ $A_{N2}$ ...) away from the diagonal can usually be approximated by low-rank matrices.

In the context of the 3D Helmholtz equation, the hierarchical decomposition provided by the H-matrix reduces the computational complexity from $O(N^2)$ to $O(Nk\log N)$, where $k$ is the rank of the approximation. This reduction significantly decreases both computational complexity and storage requirements, making the SBM method more efficient and scalable for large-scale problems.

By integrating H-matrices with SBM, the dense matrices resulting from the collocation process are decomposed into hierarchical sub-blocks, allowing far-field interactions to be approximated by low-rank matrices. This approach not only reduces computational and storage demands but also enhances the efficiency of matrix operations such as matrix-vector multiplication and inversion, which are critical in solving large-scale linear systems.

## 3 NUMERICAL RESULTS

This section presents three numerical examples to demonstrate the applicability and effectiveness of the proposed meshless collocation method based on H-matrices for solving the Laplace and Helmholtz equations. To quantify the accuracy of the numerical solutions, we introduce the $L_2$ norm relative error $L_{2err}(u)$ and the compression ratio. The $L_2$ norm relative error is defined as:

$$L_{2err}(u) = max\left(\frac{\sqrt{\sum_{i=1}^{N}|\hat{u}_i-u_i|^2}}{\sqrt{\sum_{i=1}^{N}|u_i|^2}}\right). \tag{7}$$

The compression ratio is given by:

$$\text{Compression ratio} = \frac{\text{size of the original matrix}}{\text{size of the compressed matrix}}. \tag{8}$$

### 3.1 Laplace equation in a unit circle domain (2D)

The first example considers the Laplace equation within a 2D unit circle domain:

$$\begin{cases} \Delta u = 0, & \text{in } \Omega \\ u = x^2 - y^2, & \text{on } \partial\Omega \end{cases}$$

The numerical results are shown in Table 1, which indicates that $L_2$ error increases with the increase of the number of nodes. However, it is important to note that the $L_2$ error remains within acceptable limits. The compression rate increases significantly with the increase of the number of nodes, indicating that the method effectively reduces the storage requirement while maintaining acceptable accuracy.

Table 1: H-matrix for the solution of the 2D Laplace equation with different node numbers (*N*).

| *N* | **Time (sec.)** | $L_2$ **error** | **Compression ratio** | **Size of the compressed matrix (MB)** |
|---|---|---|---|---|
| 100,000 | 79.68 | 2.07E-07 | 147.72 | 1,044.26 |
| 500,000 | 356.33 | 3.36E-06 | 703.07 | 5,497.76 |
| 1,000,000 | 670.74 | 5.95E-06 | 1,324.74 | 11,662.28 |
| 1,500,000 | 1,106.58 | 3.98E-05 | 1,766.98 | 19,604.49 |
| 2,000,000 | 2,475.09 | 3.99E-05 | 2,493.87 | 24,762.15 |

### 3.2 Laplace equation in a unit sphere domain (3D)

The second example considers the Laplace equation within a 3D unit sphere domain:

$$\begin{cases} \Delta u = 0, & \text{in } \Omega \\ u = e^x \cos(\frac{\sqrt{2}}{2}y)\cos(\frac{\sqrt{2}}{2}z), & \text{on } \partial\Omega \end{cases}$$

Fig. 2 illustrates the relationship between the number of nodes $N$, the $L_2$ error, and the compression ratio for the 3D Laplace equation. The graph demonstrates that the $L_2$ error decreases with an increasing number of nodes, but beyond a certain threshold, the error slightly increases. However, it is important to note that the $L_2$ error remains within acceptable limits. Nonetheless, the compression ratio continues to improve, emphasising the method's capability to handle large-scale problems efficiently.

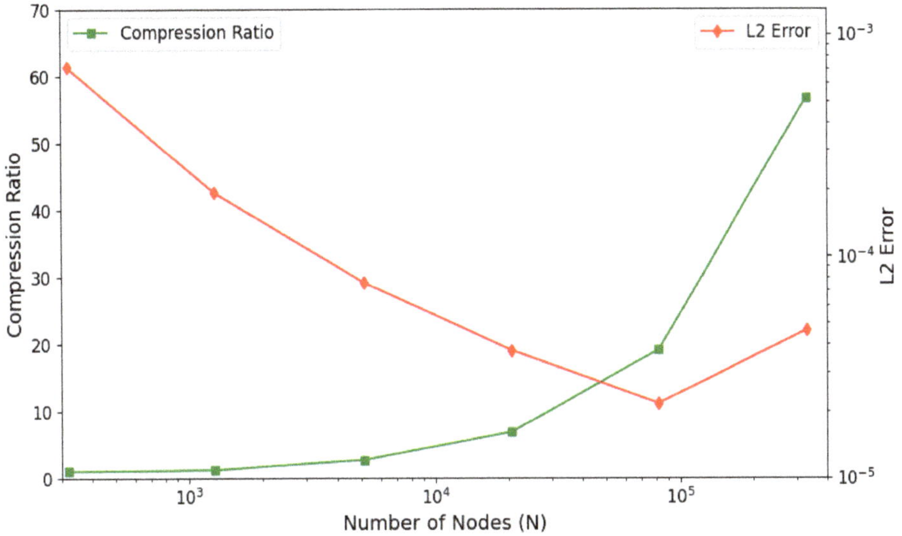


Figure 2: The numerical error of the compression rate varies with the number of collocation nodes in the 3D Laplace equation.

### 3.3 Helmholtz equation in a unit sphere domain (3D)

The third example addresses the Helmholtz equation in the same 3D unit sphere domain with a wave number $\lambda$ =1:

$$\begin{cases}(\Delta + \lambda^2)u = 0, & \text{in } \Omega \\ u = \exp(i\lambda r)/r, & \text{on } \partial\Omega\end{cases}$$

Fig. 3 presents a plot of the $L_2$ error and compression ratio against the number of nodes $N$ for this problem. The results for the Helmholtz equation are similar to those observed for the Laplace equation. The $L_2$ error decreases as the number of nodes increases, although it eventually increases slightly. However, it is important to note that the $L_2$ error remains within acceptable limits. The compression ratio improves with more nodes, reflecting the method's capability to efficiently compress the system matrix while solving the Helmholtz equation.

Next, we examine the performance of the H-matrix method for solving the 3D Helmholtz equation across various wave numbers ($\lambda$). Table 2 presents the computational time, $L_2$ error, compression ratio, and the size of the compressed matrix for different values of $\lambda$.

The results show that the computational accuracy tends to decrease significantly as the wave number λ increases. The decrease in accuracy can be attributed to the inherent oscillatory nature of the fundamental solution of the Helmholtz equation, which is particularly evident at higher wave numbers. Similarly, the compression ratio decreases with increasing wave number. This decrease suggests that the effectiveness of the method in compressing the matrix decreases as the wave number increases. In conclusion, while higher wave numbers pose a greater numerical challenge.

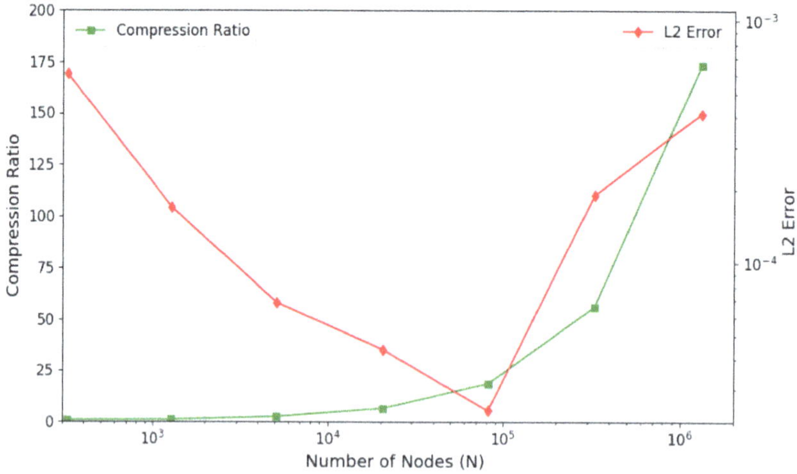


Figure 3: The numerical error of the compression rate varies with the number of collocation nodes in the 3D Helmholtz equation.

Table 2: H-matrix for the solution of the 3D Helmholtz equation with different wave numbers ($N = 327{,}680$).

| $\lambda$ | Time (sec.) | $L_2$ error | Compression ratio | Size of the compressed matrix (MB) |
|---|---|---|---|---|
| 1 | 27.99 | 4.38E-05 | 6.79 | 966.16 |
| 5 | 32.58 | 1.37E-04 | 5.86 | 1,094.23 |
| 10 | 36.55 | 2.71E-04 | 5.09 | 1,259.61 |
| 20 | 46.58 | 7.59E-04 | 4.09 | 1,567.14 |
| 30 | 63.58 | 3.43E-03 | 3.44 | 1,862.79 |
| 40 | 78.29 | 1.22E-02 | 2.97 | 2,155.63 |

## 4 CONCLUSIONS

The results of this study demonstrate that the combination of H-matrices with the SBM provides an efficient numerical approach for solving large-scale linear systems. The numerical examples, particularly those involving the Laplace and Helmholtz equations, reveal that this method significantly reduces computational complexity and memory usage while maintaining a high degree of accuracy. Its capability to handle dense asymmetric matrices without sacrificing precision establishes it as a promising alternative to traditional numerical methods. However, challenges remain in addressing high wave number problems, indicating a need for further refinement of the method. Overall, this work underscores the effectiveness of the combined H-matrix and SBM approach, laying a strong foundation for future research aimed at expanding its applicability and optimising the H-matrix algorithm to enhance computational efficiency.

## ACKNOWLEDGEMENTS

The work described in this paper was supported by the National Natural Science Foundation of China (grant no. 12122205, 12372196).

## REFERENCES

[1] Takahashi, T. & Matsumoto, T., An application of fast multipole method to isogeometric boundary element method for Laplace equation in two dimensions. *Engineering Analysis with Boundary Elements*, **36**(12), pp. 1766–1775, 2012.

[2] Aimi, A., Calabro, F., Diligenti, M., Sampoli, M.L., Sangalli, G. & Sestini, A., Efficient assembly based on B-spline tailored quadrature rules for the IgA-SGBEM. *Computer Methods in Applied Mechanics and Engineering*, **331**, pp. 327–342, 2018.

[3] Fu, Z.J. et al., Singular boundary method: A review and computer implementation aspects. *Engineering Analysis with Boundary Elements*, **147**, pp. 231–266.

[4] Chen, W. & Wang, F.Z., A method of fundamental solutions without fictitious boundary. *Engineering Analysis with Boundary Elements*, **34**(5), pp. 530–532, 2010.

[5] Chen, W., Fu, Z.J. & Wei, X., Potential problems by singular boundary method satisfying moment condition. *Computer Modeling in Engineering and Sciences*, **54**(1), pp. 65–85, 2009.

[6] Liu, Y.J. et al., Recent advances and emerging applications of the boundary element method. *Applied Mechanics Reviews*, **64**(3), 030802, 2012.

[7] Cheng, A.H.D. & Cheng, D.T., Heritage and early history of the boundary element method. *Engineering Analysis with Boundary Elements*, **29**(3), pp. 268–302, 2005.

[8] Chen, J.T., Yang, J.L., Lee, Y.T. & Chang, Y.L., Formulation of the MFS for the two-dimensional Laplace equation with an added constant and constraint. *Engineering Analysis with Boundary Elements*, **46**, pp. 96–107, 2014.

[9] Chaillat, S., Desiderio, L. & Ciarlet, P., Theory and implementation of H-matrix based iterative and direct solvers for Helmholtz and elastodynamic oscillatory kernels. *Journal of Computational Physics*, **351**, pp. 165–186, 2017.

[10] Hackbusch, W., *Hierarchical Matrices: Algorithms and Analysis*, Springer Series in Computational Mathematics, **49**, 2015.

[11] Börm, S., Grasedyck, L. & Hackbusch, W., Introduction to hierarchical matrices with applications. *Engineering Analysis with Boundary Elements*, **27**(5), pp. 405–422, 2003.

[12] Wei, X., Chen, W., Sun, L.L. & Chen, B., A simple accurate formula evaluating origin intensity factor in singular boundary method for two-dimensional potential problems with Dirichlet boundary. *Engineering Analysis with Boundary Elements*, **58**, pp. 151–165, 2015.

[13] Chen, W., Fu, Z.J., Wei, X., Potential problems by singular boundary method satisfying moment condition. *CMES-Computer Modeling in Engineering and Sciences*, **54**(1), pp. 65–85, 2009.

[14] Sun, L.L., Chen, W. & Cheng, A.H.D., Evaluating the origin intensity factor in the singular boundary method for three-dimensional Dirichlet problems. *Advances in Applied Mathematics and Mechanics*, **9**(6), pp. 1289–1311, 2017.

[15] Fu, Z.J., Chen, W., Wen, P.H. & Zhang, C.Z., Singular boundary method for wave propagation analysis in periodic structures. *Journal of Sound and Vibration*, **425**, pp. 170–188, 2018.

[16] Sun, L.L., Chen, W. & Zhang, C.Z., A new formulation of regularized meshless method applied to interior and exterior anisotropic potential problems. *Applied Mathematical Modelling*, **37**, pp. 7452–7464, 2013.

# SECTION 3
# FLUID AND SOLID MECHANICS

# SOLID–SOLID INTERFACIAL DYNAMICS AND MODELLING

WEI CHEN, LITENG YANG, LIMEI CAO & XINGHUI SI
University of Science and Technology Beijing, China

ABSTRACT
Solid interface modelling is critical to the stability, stiffness, deformation and thermal performance of a structural system. Different from interior elements, a boundary element may experience changes in material properties, loads and dynamic modes. To date, the computational results from boundary elements are less satisfactory, and the strategy to overcome that drawback is to increase the mesh density near the boundary elements or significantly reduce the time steps when dynamic loads are applied. Recently, new insight into physics has revealed that a solid–solid interface may experience nondissipative dynamics and produce nodes, antinodes and saddles dynamically. Instead of increasing the mesh density and time step, we propose explicitly including the physics of nondissipative dynamics in the finite element models to reduce the overall demands and cost of computing. In this work, we present the differential equations of nondissipative dynamics and mathematical formulas for computing nodes, antinodes and saddles, which are readily coded in any finite element software. We will show the details of the derivations of those formulas and how they can be included in the finite element models. We will also show some canonical calculations for saving computational time and model sizes and compare the results with measurements and the sensitivity to the mesh density and time steps. We expect that the revelation of new physics and mathematical formulas will increase the accuracy and efficiency of simulations for industrial applications.
*Keywords: interfacial phenomena, nondissipative dynamics, solid mechanics, speed of the first sound, speed of the second sound, pulses at interface, pulse train at interface, chaos.*

## 1 INTRODUCTION

A solid–solid interface is a basic mechanical structure, which can be found in almost every engineering application of mechanical structures. Hence, the accurate prediction of stress field, elastic variation and plastic deformation on an interface is important to the design, manufacture and operation of a machinery. To carry out the computational modelling, a numerical or computational scheme such as the finite element method (FEM) and the boundary element method (BEM) has to be chosen and the corresponding algorithm has to be coded and executed in a computational resource. Although the stress field are readily resolvable inside the solid domains, the interfacial damages and deformation have shown the substantial discrepancies between the theoretical prediction and the lab observations. In order to resolve the disparity, the significant effort has been made to develop sophisticated numerical schemes such as BEM. As the result, BEM is particularly suited for the interfacial phenomena because only discretisation of surfaces is necessary; hence, it can adapt to any complexity of surface topologies in a contact. Additionally, the sensitive, oscillatory differential equations at an interface can be applied and solved with high precision algorithms [1]. However, due to the assumptions made on the integral equations and the coupling of the normal and tangential components of stress and strain at the contact, the calculation of plastic deformation, sliding and strain rates still deviates from observations [2]. Recently, the theory on the isentropic motion in solids has shed some lights on the hydrodynamic anomalies in the motion of solids. One of outcomes from the theory of excess entropy or isentropic motion is the interfacial effects, which means that under the oscillatory loads, the interface resonates and generates sound. Acoustic propagation creates a discontinuity of field variables across

WIT Transactions on Engineering Sciences, Vol 136, © 2024 WIT Press
www.witpress.com, ISSN 1743-3533 (on-line)
doi:10.2495/BE470081

an interface. The interfacial phenomena are strongly dependent on this nondissipative dynamics. In this work, we give an introduction on this new theory and present the validation by comparing it with the experimental data. Through the mathematical derivation and experimental data validation, we hope to supply the formulas that can be used in the calculation of strain and stress field without additional burden on the finite element method and computation.

The solid–solid interface has been considered as the multiscale problem [3], [4]. For the historical reasons, the interfacial problems have been dealt with each discipline such as heat transfer, mechanical and electrical issues [3], [5]–[7]. The essence of the theoretical work has been more or less focused on the geometrical distribution of the contact. When two nominally perfect surfaces are in contact, the interface is assumed to be separated by many geometrical spikes and non-uniformities such that the contact is only partially in physical contact. These protrusions have random shapes and sizes and the stress and strain on the interface therefore depends on the distribution statistically [8]. Such treatment of contact surfaces has been known to employee a parameter called roughness and it has been found that the roughness alone is not sufficient to describe the mechanical stiffness for the texture effects [9]. The thermal performance of an interface has been related to the so-called interface thermal conductance where an additional temperature step has been observed. This phenomenon has been called Kapitza conductance to celebrate his discovery in the early 20th century [10]. Initially, the interface thermal conductance has been thought as the phenomenon particularly for Helium at low temperature [11]. Later, the interfacial temperature differences have been found in many solid–solid interfaces and the connection to the speed of sound has been established [6]. The interfacial electric conductance has also been observed and another phenomenon associated with the interfacial conductance is the so-called tunnelling effect [12]. Although the thermal-electron tunnelling phenomena have been observed over a wide range of materials, there is few on the possible tunnelling phenomenon in the mechanical behaviour [13], where the tunnelling effect has been studied by the quantum effect while the interfacial effect is not. Overall, while some significant progress on the interfacial phenomena has been made in applications, a coherent theory and mathematical models are not yet available to consolidate various theories and models.

When the finite element method becomes available, the expectation was high to address the interfacial issues [14]; however, the challenges faced by FEM left an opportunity for the development of BEM [1], [8], [15], [16]. The principle of BEM is to 'discretising' a volumetric domain into matrices and fitting the exact boundary conditions to the domain matrices through integral equations casted by the Green's functions, which is advantageous over other computational methods on the accuracy of the solutions on the boundaries [17]. However, the applications and experiences have exposed some shortcomings and difficulties at interfaces [2], [18]–[20].

The interfacial mechanical property is an old scientific field dated back to da Vinci era [21]. The interfacial science relates to many subjects such as physics, chemistry, material science, biology, and engineering. The pioneers of tribology rely on the concept of friction from the roughness or asperity of interfaces [22]. In the classical tribology, the work to properly account for the loss of energy and increase of resistance has been weighted on the interpretation of distribution of real contact areas. The direct consequences of the friction are then the wearing of contact areas accompanied with the sound [23]. In addition to the plastic deformation and acoustic emission, the researches on tribology have revealed that the responses of the sliding interfaces are strong function of time [24], [25]. The so-called rigid modes are included on the interaction of the primary structure motions with the non-classically damped modal properties [26].

It is almost impossible to judge whether the advances in the theoretical fronts and the meshing techniques have unlocked the mystery of interfacial phenomena. However, one thing is clear that the developments, regardless of the theoretical, numerical or experimental work, are exploratory and fragmented in nature. In other words, we may have not yet lodged on a coherent physics of interfacial phenomena. Recently, a study on the nondissipative dynamics hypothesises that in any periodic motion, the dissipative and nondissipative dynamics coexists with their own laws of motion [27]. With that theory, the temperature discontinuity at interfaces has been associated with the speed of the first and the second sound. Moreover, the predicted temperature profile on the sites of excitation at the interface has been compared to the experimental measurements. The theory assumes that in one period, there shall be at least one point in time that entropy in the zero. At that particular point, we must employee the governing equations for nondissipative motion instead of the dissipative process. When a solid pocket follows the nondissipative dynamics, many new behaviours and physics are exposed, which may lead to the resolution of the interfacial phenomena.

In this work, we follow the approach of nondissipative dynamics, formulate the motion of solids at the interface and validate it with the experimental data. The formulas can be used to calculate the interfacial plastic deformation and friction resistance incurred. The article is organised in the following order. First, we separate the nondissipative motion from the dissipative motion in an isotropic, Hookean solid. Second, we establish the boundary conditions for the nondissipative motions of solids. Then, we solve the velocity on a two-dimensional interface and validate it with the experimental data. We end our study with a short conclusion.

## 2 ISENTROPIC AND ENTROPIC MOTIONS

In this section, we develop a means to separate nondissipative motion from the dissipative motion.

The purpose is that the isentropic motion follows the different physics of separating the isentropic motion from the entropic motion. In the modern theoretical physics and fluid mechanics, a nondissipative motion is ideal and does not exist in reality; however, in the theory of isentropic dynamics, the nondissipative motion actually exist and represents a different thermodynamic state and there is a conversion between entropic solid and isentropic solid. When an oscillatory load is applied, an isentropic motion can excite the domain of interest and result in the resonance and generate sounds. On the contrary, a theoretical nondissipative solid never goes beyond the limit of an entropic solid; moreover, the nondissipative solid has little to do with acoustics and supersolid that we will discuss with next.

In this work, we start with the governing equations of isotropic, linear, elastic, Hookean solid in three dimensional motions in spatiotemporal [28], [29]:

$$\frac{D\rho}{Dt} + \rho\, \nabla \cdot \mathbf{u} = 0 \tag{1}$$

$$\rho \frac{D\mathbf{u}}{Dt} = \nabla \cdot \Xi \tag{2}$$

$$\rho \frac{Dh}{Dt} = \Xi : \Lambda - \nabla \cdot \mathbf{q} \tag{3}$$

where $\rho$, $h$, $\mathbf{u}$, $\mathbf{q}$, $\Xi$ and $\Lambda$ are the density, enthalpy, velocity and heat flux vectors, stress and rate of deformation tensors, respectively. Here, we note the equivalence of the local and instant velocity gradient, strain rate and the rate of deformation. Consequently, the stress tensor is followed by the constitutive relation of the Hooke's law,

$$\Xi_{ij} = -p\delta_{ij} + \lambda\mathcal{E}_{kk}\delta_{ij} + 2\mu\mathcal{E}_{ij} \tag{4}$$

where $p$ is the hydrostatic pressure, $\delta_{ij}$ is the Kronecker symbol, $\mathcal{E}_{ij}$ is the strain deviator tensor and $\lambda$ and $\mu$ are the Lame constants, which relate to the more familiar shear modulus G, Young's modulus E and Poisson ratio $\upsilon$ by

$$\mu = G = \frac{E}{2(1+\upsilon)} \quad \text{and} \quad \lambda = \frac{\upsilon E}{(1+\upsilon)(1-2\upsilon)} \tag{5}$$

The rate of deformation tensor can be expressed by

$$\Lambda = \Lambda_{ij} = \frac{1}{2}\left(\frac{\partial u_i}{\partial x_j} + \frac{\partial u_j}{\partial x_i}\right) \tag{6}$$

For an isotropic, elastic solid, we adopt the Venant–Beltrami theorem [29] and find that the deviators of the strain tensor $\mathcal{E}_{ij}$ are traceless, i.e. the normal stress $\mathcal{E}_{kk} = 0$ [29]. The strain deviator tensor $\mathcal{E} = \mathcal{E}_{ij}$ then follows by

$$\mathcal{E} = \mathcal{E}_{ij} = \frac{1}{2}\left(\frac{\partial \xi_i}{\partial x_j} + \frac{\partial \xi_j}{\partial x_i}\right) \; i \neq j \tag{7}$$

From the thermodynamic relation,

$$h = e + \frac{p}{\rho} \tag{8}$$

where $e$ is the specific mass density of internal energy. Bring (4), (6) and (8) into (3) and utilise (1),

$$\rho\frac{De}{Dt} = -pI\colon\Lambda + 2\mu\mathcal{E}\colon\Lambda - \nabla\cdot\mathbf{q} \tag{9}$$

From the thermodynamic relation of the internal energy on temperature,

$$\rho\frac{De}{Dt} = \rho c_v\frac{DT}{Dt} + \left[T\left(\frac{\partial p}{\partial T}\right)_\rho - p\right]\frac{\partial u_i}{\partial x_i} \tag{10}$$

(10) can be expressed by temperature,

$$\rho c_v\frac{DT}{Dt} = -pI\colon\Lambda + 2\mu\mathcal{E}\colon\Lambda - \nabla\cdot\mathbf{q} + \left[p - T\left(\frac{\partial p}{\partial T}\right)_\rho\right]\frac{\partial u_i}{\partial x_i} \tag{11}$$

The thermodynamic relation of the internal energy on temperature, entropy, pressure and specific volume yields

$$de = Tds + pdv \tag{12}$$

or

$$\frac{De}{Dt} = T\frac{Ds}{Dt} + p\frac{Dv}{Dt} = T\frac{Ds}{Dt} - \frac{p}{\rho^2}\frac{D\rho}{Dt} \tag{13}$$

With (13), (11) becomes

$$\rho T\frac{Ds}{Dt} - \frac{p}{\rho}\frac{D\rho}{Dt} = -p\frac{\partial u_i}{\partial x_i} + 2\mu\mathcal{E}\colon\Lambda - \nabla\cdot\mathbf{q} \tag{14}$$

With the continuity eqn (1), the energy conservation is expressed by the local mass density of entropy,

$$\rho T\frac{Ds}{Dt} = 2\mu\mathcal{E}\colon\Lambda - \nabla\cdot\mathbf{q} \tag{15}$$

Here, we introduce a new variable, excess entropy $\delta s$

$$\delta s = s_0 - s \tag{16}$$

where $s$ and $s_0$ are the local mass density of the solid and the mean entropy of the domain of interest. Since it is a constant, we may assume that $s_0$ is the entropy of the nondissipative process, and when $s = s_0$, the local solid parcel experiences a nondissipative process isentropically. The excess entropy divides a solid domain into two thermodynamic domains: the entropic or dissipative domain and the isentropic or nondissipative domain. Excess entropy is a local quantity and can be determined by (15)

$$\rho T \frac{D\delta s}{Dt} = \nabla \cdot \mathbf{q} - 2\mu \mathcal{E} : \Lambda \tag{17}$$

When $\nabla \cdot \mathbf{q} = 0$ and $2\mu\mathcal{E}:\Lambda = 0$, (17) says

$$\frac{D\delta s}{Dt} = 0 \text{ or } \delta s = 0 \tag{18}$$

Note that we replace $\delta s = C$ by an integral constant $C$ to keep $\delta s = 0$. In other words, we have established criterion (18) to discern local solid on whether it is a regular solid or supersolid. The compliance of the excess entropy to the Clausius–Duhem's law is given in Chen [30].

It is necessary to explain the physical importance of (18) to the momentum eqn (2). For example, the functions of the strain tensor and heat flux depend on time by

$$\mathcal{E} = \mathcal{E}_{ij} = \cos \omega t \, \hat{\mathcal{E}}_{ij} \tag{19}$$

and

$$\mathbf{q} = \cos \omega t \, \hat{\mathbf{q}} \tag{20}$$

When $\cos \omega t = 0$, from (19) and (20), (17) yields

$$\frac{D\delta s}{Dt} = 0 \tag{21}$$

and the momentum conservation (2) yields

$$\rho \frac{D\mathbf{u}}{Dt} = -\nabla \cdot p \tag{22}$$

Therefore, we conclude that the governing equations for a nondissipative solid have the form of

$$\frac{D\rho}{Dt} + \rho \nabla \cdot \mathbf{u} = 0 \tag{23}$$

$$\rho \frac{D\mathbf{u}}{Dt} = -\nabla \cdot p \tag{24}$$

$$\rho \frac{Dh}{Dt} = 0 \tag{25}$$

It is clear that the governing eqns (23)–(25) of a nondissipative solid or supersolid are identical to the governing equations of the superfluid [31], as suggested [32], [33]. From the Maxwell relations in thermodynamics,

$$dp = \left(\frac{\partial p}{\partial \rho}\right)_s d\rho \tag{26}$$

Apply (26) to (23) and (24), and take the derivative with respect to the time, we have

$$\frac{1}{\left(\frac{\partial p}{\partial \rho}\right)_s}\frac{\partial p}{\partial t} + \nabla \cdot (\rho \mathbf{u}) = 0 \tag{27}$$

Take the derivative of (27) with respect to time,

$$\frac{1}{\left(\frac{\partial p}{\partial \rho}\right)_s}\frac{\partial^2 p}{\partial t^2} + \nabla \cdot \left[\rho \frac{\partial (\mathbf{u})}{\partial t}\right] - \nabla \cdot [\mathbf{u}\nabla \cdot (\rho \mathbf{u})] = 0 \tag{28}$$

Take the divergence to (24), subtract it from (28), and rearrange,

$$\frac{\partial^2 p}{\partial t^2} - \left(\frac{\partial p}{\partial \rho}\right)_s \nabla^2 p = \Psi \tag{29}$$

where $\Psi = \left(\frac{\partial p}{\partial \rho}\right)_s \nabla \cdot (\rho \mathbf{u} \nabla \cdot \mathbf{u})$ is the nonlinear term. (29) is the wave equation for the momentum conservation of supersolid.

Take the derivative to (25) with respect to $\mathbf{x}$ and $t$ and subtract the former from the latter,

$$\frac{\partial^2 T}{\partial t^2} - \mathbf{u} \cdot \mathbf{u} \nabla^2 T = \Phi \tag{30}$$

where $\Phi = -\frac{1}{\rho}\frac{\tilde{D}\rho}{\tilde{D}t}\left(\frac{\partial T}{\partial t} + \mathbf{u} \cdot \nabla T\right) + \frac{\tilde{D}\mathbf{u}}{\tilde{D}t}\nabla T - \frac{\beta}{\rho}\frac{\tilde{D}T}{\tilde{D}t}\nabla \cdot \mathbf{u} + \frac{\beta T}{\rho}\frac{\tilde{D}}{\tilde{D}t}(\nabla \cdot \mathbf{u})$ and $\beta = -\frac{1}{\hat{C}_V}\left(\frac{\partial p}{\partial T}\right)_V$. (30) is the wave equation for the energy conservation of supersolid.

From (29) and (30), we can construct a two-dimensional dynamical system $(p, T)$ that

$$F(p,T) = \frac{\partial^2 p}{\partial t^2} - \left(\frac{\partial p}{\partial \rho}\right)_s \nabla^2 p \tag{31}$$

$$G(p,T) = \frac{\partial^2 T}{\partial t^2} - u^2 \nabla^2 T \tag{32}$$

At any arbitrary wavelets,

$$p = p_0 \sin(\mathbf{x} + \mathbf{c_1} t) \text{ and } T = T_0 \sin(\mathbf{x} + \mathbf{c_2} t) \tag{33}$$

where $c_1$ and $c_2$ are the characteristic speeds of the momentum and energy disturbances.

When the dynamical system experiences the disturbances of (33), (29) and (30) produce

$$F(p,T) = \frac{\partial^2 p}{\partial t^2} - \left(\frac{\partial p}{\partial \rho}\right)_s \nabla^2 p = -\left[c_1^2 - \left(\frac{\partial p}{\partial \rho}\right)_s\right] p = \Psi \tag{34}$$

$$G(p,T) = \frac{\partial^2 T}{\partial t^2} - u^2 \nabla^2 T = -[c_2^2 - u^2] T = \Phi \tag{35}$$

The nullified Jacobian of the dynamical system (34) and (35) yields the condition of resonance,

$$J = \begin{bmatrix} \frac{\partial F}{\partial p} & \frac{\partial F}{\partial T} \\ \frac{\partial G}{\partial p} & \frac{\partial G}{\partial T} \end{bmatrix} = -\begin{bmatrix} c_1^2 - \left(\frac{\partial p}{\partial \rho}\right)_s & 0 \\ 0 & c_2^2 - u^2 \end{bmatrix} = 0 \tag{36}$$

The condition for resonance of superfluid is given by (36) and calculated by Chen [34]

$$c_1^2 = \left(\frac{\partial p}{\partial \rho}\right)_s = \frac{E}{\rho} \tag{37}$$

and from $dT = \left(\frac{\partial T}{\partial \rho}\right)_s d\rho$ for an isentropic process,

$$c_2^2 = -\frac{R\rho}{M}\left(\frac{\partial T}{\partial \rho}\right)_S = \frac{R_S\alpha_p T}{\lambda_0\kappa_T c_p\rho + \alpha_p^2 T} \tag{38}$$

where $c_1$ is the speed of the first sound and $c_2$ is the speed of the second sound. $R_s = \frac{R}{M}$ is the specific gas constant, $\kappa_T$ is the isothermal compressibility, $\alpha_p$ is the isochoric thermal expansion coefficient, $c_p$ is the isobaric specific heat and $\lambda_0 = 10^{-3}$. Some validation on the speed of the second sound can be found in Chen [27]. Another important result from the analysis of governing equations of the resonant solids are written as

$$\begin{cases} \rho\left(\frac{\partial \boldsymbol{u}}{\partial t} \pm \boldsymbol{c_1} \cdot \nabla \mathbf{u}\right) = -\nabla p \\ \rho\left(\frac{\partial \mathrm{T}}{\partial t} \pm \boldsymbol{c_2} \cdot \nabla \mathrm{T}\right) = 0 \end{cases} \tag{39}$$

Now, we have established the governing differential equations of supersolid (23), (24) and (25), resonant supersolid (39) from the governing differential equations of a Hookean solid (1), (2) and (3).

## 3 BOUNDARY OF THE SECOND LAW

In this section, we establish the initial and boundary conditions for the isentropic motions of a solid that are used to specify the integral constants in the solutions of supersolid and resonant supersolid.

From the theory of nondissipative dynamics, a regular solid and supersolid coexist dynamically and are mutually dependent and what connects and apportions them is the endless oscillatory motion; on the other hands, the existence of regular solid and supersolid sustains a perpetual motion. Therefore, the motion in regular solid activates supersolid and the velocity and temperature pass the momentum and energy from the regular solid to supersolid. The passing of mass, momentum and energy to supersolid can be described by the initial and boundary conditions on the boundary of the second law. These boundary conditions can be used to determine the coefficients from the general solutions of velocity and temperature in (39).

Quantitatively, when (16) is satisfied, the thermodynamic state of solid changes from entropic to isentropic, the governing equations are also changed from (1), (2) and (3) to (23), (24) and (25) or (39). This change results in the sudden impulses in solid because there is no impedance term in the governing equations of supersolid and the amplitudes of velocity and temperature on the side of regular solid will be magnified on the side of supersolid. This train of impulses is conveniently expressed by the so-called Dirac combs. For example, if we assume that

We assume that there is a displacement perturbation on the regular solid side,

$$\xi^\dagger = \xi_0^\dagger \sin(2\pi f k_2 x) \sin(2\pi f k t) = \varphi(x,y)\alpha_k(t) \tag{40}$$

where $\xi^\dagger$ is the displacement and $\xi_0^\dagger$ is the corresponding amplitude of displacement. Superscript † is used to specify variables on the regular solid side. According to (6) and (7),

$$\Lambda = \Lambda_{ij} = \frac{1}{2}\left(\frac{\partial u_i}{\partial x_j} + \frac{\partial u_j}{\partial x_i}\right) = \pi f k\left(\frac{\partial \varphi}{\partial x} + \frac{\partial \varphi}{\partial y}\right)\cos(2\pi f k t) \tag{41}$$

$$E = E_{ij} = \frac{1}{2}\left(\frac{\partial \xi_i}{\partial x_j} + \frac{\partial \xi_j}{\partial x_i}\right) = \frac{1}{2}\left(\frac{\partial \varphi}{\partial x} + \frac{\partial \varphi}{\partial y}\right)\alpha_k(t) = \frac{1}{2}\left(\frac{\partial \varphi}{\partial x} + \frac{\partial \varphi}{\partial y}\right)\sin(2\pi f k t) \tag{42}$$

To guarantee the second term on the right of (17) is zero,

$$E:\Lambda = \frac{\pi f k}{2}\left(\frac{\partial \varphi}{\partial x} + \frac{\partial \varphi}{\partial y}\right)^2 \sin(2\pi f k t)\cos(2\pi f k t) = 0 \tag{43}$$

we shall have

$$\sin(2\pi f k t)\cos(2\pi f k t) = 0 \tag{44}$$

By assuming $f = k = 1$, we illustrate the temporal terms in (40), (42), (44) and velocity (45),

$$u^{\dagger} = \frac{d\xi^{\dagger}}{dt} = u_0^{\dagger}\cos(2\pi f k t) \tag{45}$$

where $u_0^{\dagger} = 2\pi\xi_0^{\dagger}$ and it should be noted that we apply the derivative only to one period; hence the velocity is replicated through compound frequency $fk$, not amplified. At $t = 0, 0.25, 0.5, 0.75, 1.0$, the function $\sin(2\pi f k t)\cos(2\pi f k t)$, marked by the solid thick curve, is zero and they are on the boundary of the second law and solid parcels become supersolid. At $t = 0.25\ and\ 0.75$, the corresponding velocity of the solid parcel is zero; then, the velocity of solid passing onto supersolid is null. At $t = 0, 0.5, 1.0$, the corresponding velocity of the regular solid parcels are at the extrema, marked by the thick arrowed lines, and at $t = 0, 1.0$, the velocity has the maximum and at $t = 0.5$, the velocity has the minimum.

At $t = 0, 0.5, 1.0$, the solid becomes supersolid and the governing equations are changed from (1), (2), (3) to (39) and supersolid experiences velocity impulses $+u_0^{\dagger}$ and $-u_0^{\dagger}$, which are given in the Dirac combs

$$\pitchfork_1 (t) = u_0^{\dagger} \Sigma_1^K \delta(t - k + 0) \tag{46}$$

$$\pitchfork_2 (t) = -u_0^{\dagger} \Sigma_1^K \delta(t - k + 0.5) \tag{47}$$

where $\pitchfork_1$ and $\pitchfork_2$ represent the Dirac comb in positive impulses and negative impulses by the dark lines with arrows in Fig. 1. At a given frequency

$$f = \frac{k}{\lambda} \tag{48}$$

where $k$ is the wavenumber and $\lambda$ is the period. The Fourier transformation of (66) and (67) with the spacing $\frac{2\pi\lambda}{k}$,

$$\pitchfork_1 (k) = \frac{1}{2\pi} u_0^{\dagger} \Sigma_{k=1}^K f\delta(k) \tag{49}$$

$$\pitchfork_2 (k) = -\frac{1}{2\pi} u_0^{\dagger} \Sigma_{k=1}^K f\delta(k) \tag{50}$$

which yields the Fourier coefficients for the Dirac combs,

$$s_1(k) = \frac{1}{2\pi} u_0^{\dagger} f \tag{51}$$

$$s_2(k) = -\frac{1}{2\pi} u_0^{\dagger} f \tag{52}$$

Now, we work on the boundary conditions. For simplicity, we assume a one-dimensional problem as indicated by (40). Recall that to satisfy the conditions that the excess entropy becomes zero by (16), (43) must be guaranteed or

$$\left(\frac{\partial \varphi}{\partial x}\right)^2 \sin(2\pi f k t)\cos(2\pi f k t) = 0 \tag{53}$$

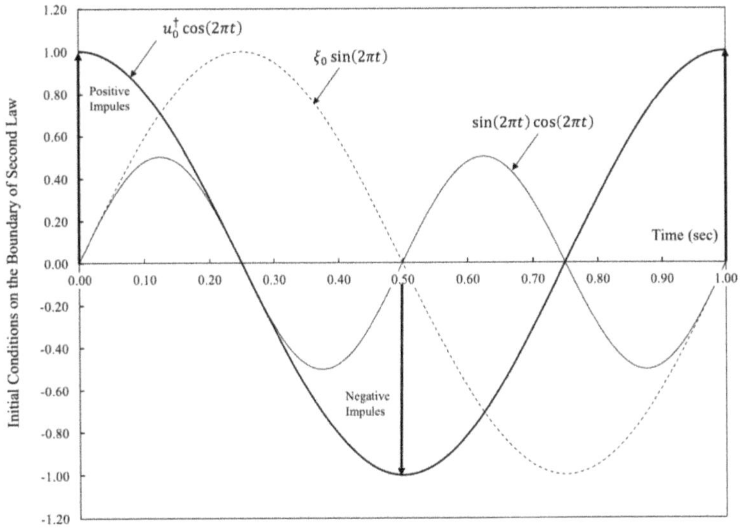


Figure 1: Initial conditions on the boundary of the second law.

We have discussed the temporal conditions for the initial conditions of the part in (53), $\sin(2\pi fkt)\cos(2\pi fkt)$. The velocity can be expressed by the spatial distribution from (45),

$$u^\dagger = \frac{d\xi^\dagger}{dt} = u_0^\dagger \sin(2\pi f k_2 x)\cos(2\pi fkt) \tag{54}$$

The spatial function $\left(\frac{\partial\varphi}{\partial x}\right)^2$ can also be nullified just like the temporal component from (53),

$$\left(\frac{\partial\varphi}{\partial x}\right)^2 = [2\pi f k_2 \cos(2\pi f k_2 x)]^2 = 0 \tag{55}$$

The physical meaning of (54) and (55) can be visualised in Fig. 2. For convenience, we divided the spatial dimension by a nominal dimension to keep $x$ dimensionless and assume a constant period of $2\pi$. There are two curves representing the velocity (54) and dissipation term in the excess entropy eqn (55). At $x = 0.25, 0.75$, there are two points that the dissipation spatial term intercepts with the horizontal axis, which means that at these two points, the excess entropy is zero and locally, the nondissipative thermodynamic state has been observed. At $x = 0.25, 0.75$, the resonant solid is adjacent to the regular solid, which shall hold the velocity quite close to the peak of the space if we are only concerned with the local velocity. This resonant solid parcel surrounded by the regular solid; hence the velocity of the regular solid is used to assign the boundary conditions of the resonant supersolid at $x = 0.25, 0.75$. From Fig. 2, we can see that at $x = 0.25$, the velocity is positive while at $x = 0.75$, the velocity is negative. If we consider continuously the displacement function (40), the positive and negative impulses can be treated with the Dirac combs as we have done to the initial conditions (46) and (47).

$$\dot{m}_{1x}(x) = u_0^\dagger \textstyle\sum_1^{K_2} \delta(x - k_2 + 0.25) \tag{56}$$

$$\dot{m}_{2x}(x) = -u_0^\dagger \textstyle\sum_1^{K_2} \delta(x - k_2 + 0.75) \tag{57}$$

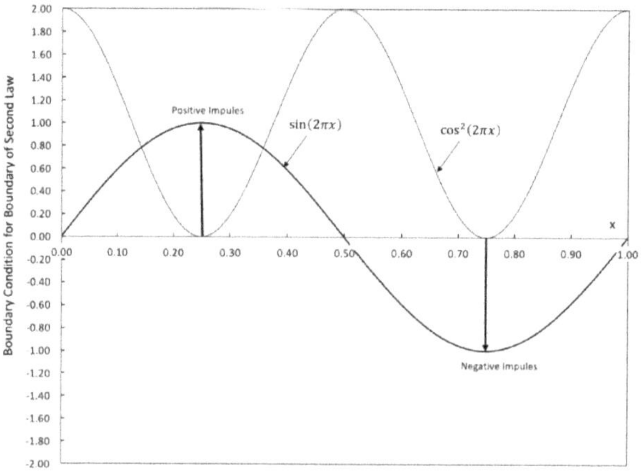


Figure 2: Boundary conditions on the boundary of the second law.

The initial conditions (46), (47) and boundary conditions (56), (57) are antinodes of the resonance and contribute to the resonance and sustain the spatiotemporal nondissipative dynamics. On the other hand, the points in the space where the velocity is zero are nodes. According to Figs 1 and 2, the nodes in space and time from the displacement function (40) are dissipative since they do not comply to the condition of the boundary of the second law. Actually, a node can be dissipative or nondissipative depending on if they are on the boundary of the second law.

In this section, we have established the initial and boundary conditions for the velocity of the resonant supersolid. In the next section, we will apply the initial and boundary conditions to determine the velocity field of the resonant supersolid.

## 4 VELOCITY DISTRIBUTION

In this section, we solve a one-dimensional spatiotemporal velocity field of resonant supersolid and the velocity field will be validated with the experimental data in the next section.

In the preceding sections, we derive the governing equations of nondissipative, resonant solids from the governing equations of dissipative, regular solid. Here, we must pause temporarily in order to clarify several critical concepts and differentiate them from the conventional meanings in theoretical physics and the theory of partial differential equations. First, the nondissipative motions here is fundamentally distinct from the nondissipative system in the theoretical physics [35], [36]. In the theory of excess entropy, the nondissipative motion is the real solid motion at the isentropic thermodynamic state and the nondissipative and dissipative solids are dependent of each other. Moreover, the real solid nondissipative motion can be excited into resonance and generate sound. Since it experiences the impulsively initial and boundary conditions, the real nondissipative process is irreversible. Finally, the real, nondissipative, resonant solid follows (39), which is completely different from that of the dissipative solid. The mathematical formulation of the nondissipative system of a solid is usually considered the degenerate partial differential equations of the partial differentiate equations of a dissipative system [37], [38]. Strongly associated with the degenerate partial differential equations, the singularity of the governing equations of an isotropic Hookean solid occurs at so-called 'vanishing dissipation' conditions [39], [40]. It is arguable that the governing differential equations of the nondissipative solid are degenerate equations at the 'vanishing dissipation' singularity. The question has been raised on whether the dissipative partial differential equation has already included the nondissipative equations; therefore, the solutions of the motions of the dissipative solids also include those of the nondissipative solids. The rebuttal on this point of view is demonstrated by the derivation and validation cases in this work. In another word, the solution of the degenerate partial differential equations of the nondissipative motions cannot be elaborated by the original partial differential equations.

In this work, we will solve one-dimensional resonant supersolid and compare it with the experimental data. From (39), the one-dimensional partial differential equation for resonant, supersolid is given below,

$$\frac{\partial u}{\partial t} + c_1 \frac{\partial u}{\partial x} = -\frac{1}{\rho}\frac{\partial p}{\partial x} \tag{58}$$

Eqn (58) is a first-order, inhomogeneous partial differential equation and can be written as

$$u = u_h + u_p \tag{59}$$

where $u_h$ and $u_p$ are the general solution of the homogeneous differential equation and the particular solution of the inhomogeneous part, respectively. Application of the separation method to the homogeneous differential eqn (59),

$$u_h = \mathrm{A(t)B(x)} \tag{60}$$

Bring (60) to (58),

$$\mathrm{B}\frac{d\mathrm{A}}{dt} + c_1\mathrm{A}\frac{d\mathrm{B}}{dx} = 0 \tag{61}$$

Divide (61) by (60),

$$\frac{1}{A}\frac{d\mathrm{A}}{dt} + \frac{c_1}{B}\frac{d\mathrm{B}}{dx} = 0 \tag{62}$$

Since each term in (62) are functions of time and space only, we can write them in the eigenfunctions,

$$n_1 + n_2 = 0 \tag{63}$$

where $n_i$, $i = 1{,}2$ are the complex eigenvalues, expressed by

$$\begin{cases} n_1 = (\alpha_1 + if_1)k_1 \\ n_2 = (\alpha_2 + if_2)k_2 \end{cases} \tag{64}$$

where $\alpha_i$, $f_i$ and $k_i$ are amplitude factors, frequency factors and harmonics of time and space for velocity respectively. From (63) and (64), $n_i$, $\alpha_i$, $f_i$ and $k_i$ are not all independent variables.

From the definition of eigenvalues (63) and (64),

$$\frac{1}{A}\frac{d\mathrm{A}}{dt} = -n_1, \frac{c_1}{B}\frac{d\mathrm{B}}{dx} = -n_2 \tag{65}$$

we have the general solutions,

$$u_h = \sum_{k_1=1}^{K_1}\sum_{k_2=1}^{K_2} A_0 B_0 e^{-(\alpha_1+if_1)k_1 t} e^{-\frac{(\alpha_2+if_2)}{c_1}k_2 x} \tag{66}$$

where $A_0$ and $B_0$ are the constants to be determined by the boundary conditions. If it is known, the pressure in (58) can be expanded into series along the kernel of the homogeneous solutions (66),

$$p = \sum_{k_1=1}^{K_1} \sum_{k_2=1}^{K_2} p_A p_B e^{-(\alpha_1+if_1)k_1 t} e^{-\frac{(\alpha_2+if_2)}{c_1}k_2 x} \tag{67}$$

By assuming the particular solution in (59) has the form of

$$u_p = \sum_{k_1=1}^{K_1} \sum_{k_2=1}^{K_2} A_p B_p e^{-(\alpha_1+if_1)k_1 t} e^{-\frac{(\alpha_2+if_2)}{c_1}k_2 x} \tag{68}$$

We can determine the coefficients $A_p$ and $B_p$ in (68) by the coefficients $p_A$ and $p_B$ through (67); otherwise, we can iterate to obtain $A_p$ and $B_p$ numerically, which implies that we can express the solution of velocity by

$$u = \sum_{k_1=1}^{K_1} \sum_{k_2=1}^{K_2} A_s B_s e^{-(\alpha_1+if_1)k_1 t} e^{-\frac{(\alpha_2+if_2)}{c_1}k_2 x} \tag{69}$$

where $A_s B_s = A_0 B_0 + A_p B_p$.

We call (69) for velocity the general solution of the resonant supersolid. When $\alpha_1 = 0$ and $\beta_1 = 0$, we obtain the quasi-steady-state solution of temperature and velocity,

$$u = \sum_{k_1=1}^{K_1} \sum_{k_2=1}^{K_2} A_s B_s e^{-if_1 k_1 t} e^{-\frac{(\alpha_2+if_2)}{c_1}k_2 x} \tag{70}$$

Therefore, the quasi-steady-state motion (70) in the resonant supersolid consists of sinusoidal and cosinusoidal waves.

Next, we rearrange (70) so that they are explicit functions of time only and prepare to apply the boundary conditions in the next section. We first look at terms in (70),

$$e^{-(\alpha_1+if_1)k_1 t} e^{-\frac{(\alpha_2+if_2)}{c_1}k_2 x}$$

$$= e^{-\alpha_1 k_1 t}(\cos f_1 k_1 t + i \sin f_1 k_1 t) e^{-\frac{\alpha_2}{c_1}k_2 x}\left(\cos\frac{f_2}{c_1}k_2 x + i \sin\frac{f_2}{c_1}k_2 x\right)$$

$$= e^{-\alpha_1 k_1 t} e^{-\frac{\alpha_2}{c_1}k_2 x}\Big(\cos f_1 k_1 t \cos\frac{f_2}{c_1}k_2 x - \sin f_1 k_1 t \sin\frac{f_2}{c_1}k_2 x + i \cos f_1 k_1 t \sin f_1 k_1 t + i \sin f_1 k_1 t \cos\frac{f_2}{c_1}k_2 x\Big) \tag{71}$$

By grouping and taking the real terms in (71), the velocity is the explicit function of time,

$$u = \sum_{k_1=1}^{K_1} \sum_{k_2=1}^{K_2} \left(\varphi_{1,k} \cos f_1 k_1 t \cos\frac{f_2}{c_1}k_2 x - \varphi_{2,k} \sin f_1 k_1 t \sin\frac{f_2}{c_1}k_2 x\right) e^{-\alpha_1 k_1 t} e^{-\frac{\alpha_2}{c_1}k_2 x} \tag{72}$$

where $\varphi_{1,k}$ and $\varphi_{2,k}$ are implicit functions of the coefficients $A_s$, $B_s$, $f_i$, $\alpha_i$ and $k_i$, $i = 1,2$. (72) can be written in dependent of $t$ and $x$ respectively,

$$u = \sum_{k_1=1}^{K_1} (\Phi_{1,k} \cos f_1 k_1 t - \Phi_{2,k} \sin f_1 k_1 t) e^{-\alpha_1 k_1 t} \tag{73}$$

and

$$u = \sum_{k_2=1}^{K_2} \left(\Omega_{1,k} \cos\frac{f_2}{c_1}k_2 x - \Omega_{2,k} \sin\frac{f_2}{c_1}k_2 x\right) e^{-\frac{\alpha_2}{c_1}k_2 x} \tag{74}$$

where $\Phi_{1,k} = \sum_{k_2=1}^{K_2} \varphi_{1,k} \cos\frac{f_2}{c_1}k_2 x\, e^{-\frac{\alpha_2}{c_1}k_2 x}$, $\Phi_{2,k} = \varphi_{2,k} \sin\frac{f_2}{c_1}k_2 x$,

$\Omega_{1,k} = \sum_{k_1=1}^{K_1} \varphi_{1,k} \cos f_1 k_1 t\, e^{-\alpha_1 k_1 t}$, and $\Omega_{2,k} = \varphi_{2,k} \sin f_1 k_1 t\, e^{-\alpha_1 k_1 t}$.

We apply the Fourier transformation to (56) and (73), we obtain

$$\frac{1}{2\pi k_1 f_1} u_0^{\dagger} = \int_0^{\infty} (\Phi_{1,k} \cos 2\pi k_1 f_1 t - \Phi_{2,k} \sin 2\pi k_1 f_1 t) e^{-2\pi\alpha_1 k_1 t} \cos 2\pi k_1 f_1 t\, dt \quad (75)$$

$$-\frac{1}{2\pi k_1 f_1} u_0^{\dagger} = \int_0^{\infty} (\Phi_{1,k} \cos 2\pi k_1 f_1 t - \Phi_{2,k} \sin 2\pi k_1 f_1 t) e^{-2\pi\alpha_1 k_1 t} \sin 2\pi k f_k t\, dt \quad (76)$$

Since $\Phi_{1,k}$ and $\Phi_{2,k}$ are not the function of time, the integration in (75) and (76) can be expressed by an algebraic equation

$$\begin{cases} \mathrm{M}_1 \Phi_{1,k} - \mathrm{M}_2 \Phi_{2,k} = \frac{1}{2\pi k_1 f_1} u_0^{\dagger} \\ \mathrm{M}_2 \Phi_{1,k} - \mathrm{M}_3 \Phi_{2,k} = -\frac{1}{2\pi k_1 f_1} u_0^{\dagger} \end{cases} \quad (77)$$

where

$$M_1 = \int_0^{\infty} \cos 2\pi k_1 f_1 t\, e^{-2\pi\alpha_1 k_1 t} \cos 2\pi \alpha_1 k_1 t\, dt = \frac{2f_1^2 + \alpha_1^2}{2\pi\alpha_1 k_1 (4f_1^2 + \alpha_1^2)}$$

$$M_2 = \int_0^{\infty} \sin 2\pi k_1 f_1 t\, e^{-2\pi k_1 f_1 t} \cos 2\pi k_1 f_1 t\, dt = \frac{2 f_1 \alpha_1}{4\pi\alpha_1 k_1 (4f_1^2 + \alpha_1^2)}$$

$$M_3 = \int_0^{\infty} \sin 2\pi k_1 f_1 t\, e^{-2\pi k_1 f_1 t} \sin 2\pi k_1 f_1 t dt = \frac{4 f_1^2}{4\pi\alpha_1 k_1 (4f_1^2 + \alpha_1^2)} \quad (78)$$

Solve (78),

$$\Phi_{1,k} = \frac{\alpha_1(\alpha_1 + 2f_1)}{f_1^2} u_0^{\dagger} \quad (79)$$

$$\Phi_{2,k} = \frac{\alpha_1[(2f_1^2 + \alpha_1^2) + f_1\alpha_1]}{f_1^3} u_0^{\dagger} \quad (80)$$

Following the same procedure, we can obtain the integral constants for the spatial dimension,

$$\Omega_{1,k} = \frac{\alpha_2(\alpha_2 + 2f_2)}{f_2^2} \quad (81)$$

$$\Omega_{2,k} = \frac{\alpha_2[(2f_2^2 + \alpha_2^2) + f_2\alpha_2]}{f_2^3} \quad (82)$$

Therefore, we obtain the velocity in one-dimensional, resonant solid,

$$u = u_0^{\dagger} \sum_{k_1=1}^{K_1} \sum_{k_2=1}^{K_2} \left( \frac{\alpha_1(\alpha_1+2f_1)}{f_1^2} \frac{\alpha_2(\alpha_2+2f_2)}{f_2^2} \cos 2\pi f_1 k_1 t \cos 2\pi \frac{f_2}{c_1} k_2 x - \frac{\alpha_1[(2f_1^2+\alpha_1^2)+f_1\alpha_1]}{f_1^3} \frac{\alpha_2[(2f_2^2+\alpha_2^2)+f_2\alpha_2]}{f_2^3} \sin 2\pi f_1 k_1 t \sin 2\pi \frac{f_2}{c_1} k_2 x \right) e^{-2\pi\alpha_1 k_1 t} e^{-2\pi \frac{\alpha_2}{c_1} k_2 x} \quad (83)$$

In summary, we have derived the one-dimensional, temporal velocity distribution for the resonant supersolid, which is the function of the amplitude of perturbation velocity, frequency and amplitude factors and harmonics in both spatial and temporal directions. We will validate the velocity field in the next section by experimental data.

## 5 VALIDATION

In this section, we proceed to validate the velocity distribution (83). The purpose of the validation is to demonstrate that at the interface of two solids, the supersolid occurs and dominates the essential behaviours at the interfaces. The velocity distribution derived is applicable to any kind of load and especially to the engineering and manufacture scenarios. Because the propensity of nondissipative processes at the interfaces, we believe that most of damages observed belong to this category, which cannot be predicted properly with mesh refinement and with dissipative motions only. On the contrary, if we include the nondissipative dynamics properly, the physics of interfaces is more comprehensive and the demand on the numerical parameters such as mesh and time steps etc. are more relaxed with higher or equivalent accuracy.

Obviously, there are vast of experimental cases on the interfacial phenomena between solids and a recent study on the ballistic heat transfer at the solid interface has applied the nondissipative dynamic theory [27]. Consequently, we may want to choose the experiments on the strain or strain rate at an interface. The experiments on the frictional sliding modes on the interface on two identical elastic plates seem to cover a wide range of parametric factors and provide both visual and quantitative observation and measurements [41]–[44].

The tests were conducted on the interface of two plates of the same kind of solid, Homalite-100, a brittle polyester resin. The plates are 76.2 mm long, 139.7 mm wide and 9.525 mm thick, which is the contact side of the plates. A nominal pressure is applied on two plates perpendicular to the interface. The impact loading from as cylindrical steel projectile of diameter 25 mm and length 51 mm, fired from a gas gun at various speed before hitting the plate. The loading wave is measured by a strain gage glued to the specimen. Dynamic photoelasticity is used to extract stress field around the interface. The photoelastic fringe patterns were recorded in real time using a high-speed camera capable of capturing 16 images at a rate of 16 images of 100 million frames. Two pairs of polariser plates are placed on either side of the Homalite plates that generate isochromatic fringes. The isochromatic fringes are interpreted to the contours of stresses $\sigma_1 - \sigma_2$, with $\sigma_1$ the maximum in-plane principal stress and $\sigma_2$the minimum in-plane principal stress. The testing results with two operating parameters, the pressure and impact velocity, are published.

At the instant of the impact, the interface is usually assumed to move slightly and then stops due to the friction. However, it has been observed that plate may slide along the interface substantially. The physics of this sliding is unclear and therefore, there is no mathematical representation on the sliding profile. However, it serves a good example to demonstrate the physics of nondissipative dynamics when we can properly present the velocity profiles since the supersolid has no resistance on shear. We use this example to illustrate that supersolid and resonant supersolid are dominate on the interface phenomena.

The contours of infringe show that there are about five different distinct modes, a crack-like mode, a pulse-like mode, a train of pulses, multiple pulses coalescing, and a pulse-like mode followed by a crack-like mode. All of these five modes can be classified by the velocity or the strain rate of stress and strain waves. The benchmark test was conducted by producing the impact on both plates and a high-speed wave is observed at the interface, which creates also a shock line defined by the first sound. This clearly indicates that the interface went into resonant supersolid mode, which leads to the understanding of the five modes. The crack-like mode is dominated by supersolid. The pulse-like mode is a soliton due to the resonant of supersolid and the resonant is very local and limited to the interface. The pulse-like train is the resonant supersolid mode that has over-spilt onto the adjacent solid and the resonance sustains for a longer duration in time and space. The multiple pulses coalescing to form a crack is when the resonance is in the transition into non-resonance mode. A pulse-like mode followed by a crack-like mode is also a transition mode from strong resonance to a broad band resonance or non-resonance mode.

The pulse-like train is probably a relative stable resonance mode. The temporal component behaves as chaos or turbulence and the spatial component visualises by the isochromatic infringes. In this work, we apply the mathematical solution (83) of the velocity field to model the pulse-like train. For a spatial distribution, only the spatial terms in (83) show up

$$u = u_0^{\dagger} \sum_{k_2=1}^{K_2} \alpha_2 \left( \frac{\alpha_2 + 2f_2}{f_2^2} \cos 2\pi f_2 k_2 x - \frac{(2f_2^2 + \alpha_2^2) + f_2 \alpha_2}{f_2^3} \sin 2\pi f_2 k_2 x \right) e^{-2\pi \alpha_2 k_2 x} \quad (84)$$

Note that for convenience, we replace $\frac{f_2}{c_1}$ and $\frac{\alpha_2}{c_1}$ by $f_2$ and $\alpha_2$. When we apply the parameters in Table 1, the agreement is satisfactory in Fig. 3. The data shows that the pulses are not generally symmetric. Since it is the in-plane velocity, the velocity in the positive axial dimension is slightly higher than that in the negative axial direction probably from the overall motion is in the positive direction. The published data is somewhat 'translated' from the isochromatic infringes by the finite element simulation and the translation may result in some losses of the subtle information. For example, near the location where the velocity becomes zero, the data shows a sharp turn to zero but (84) continues to extend into the negative for some nominal values and then comes back eventually. In the chaotic distribution, this negative region has been measured if we allow a non-zero baseline of velocity field. This non-zero velocity base is consistent to the asymmetric velocity profile as we have pointed before.

Table 1: Parameters for the velocity distribution.

| Parameter | $u_0$ (m/s) | $\alpha_2$ | $f_2$ | $k_2$ |
|---|---|---|---|---|
| Value | 3.096 | 0.210 | 0.040 | 1 |

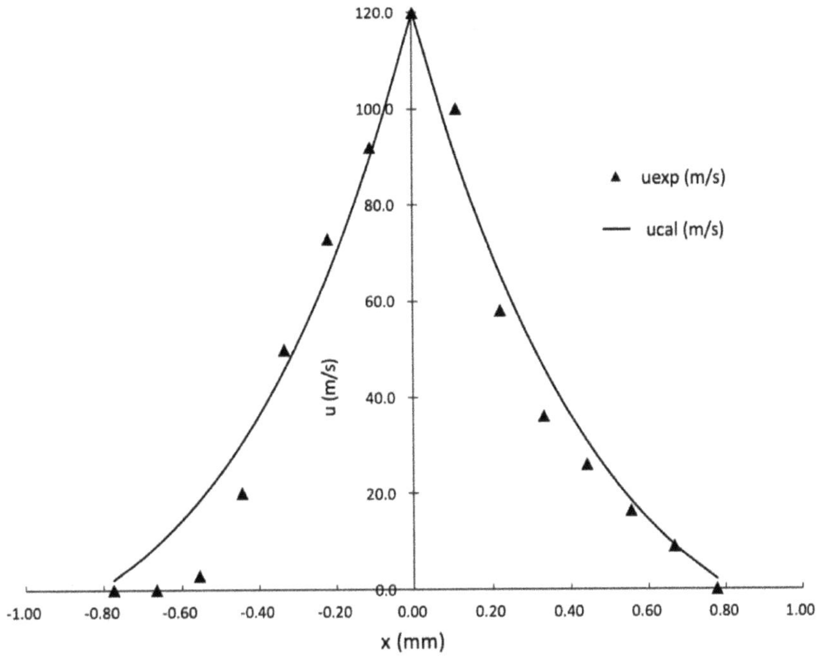


Figure 3: Experimental and calculated pulse in a pulse train at the interface.

The pulse-like train consists of a series of pulses in the space and the theory says that the temporal and spatial parts (44) and (54) of the oscillatory loads both reach zero simultaneously. The solution has temporal and spatial parts (83) and the spatial parts has the nodal and antinodal points where the velocity becomes zero or maximum and minimum. Those spatial pulses are the antinodal locations and the antinodal locations into the volumetric domains in each plate can be seen in the isochromatic infringes and the computed infringe patterns.

In Fig. 4, we plot the antinodes according to the experimental spacing. The jump from the transition of the thermodynamic states is observed from the magnitude of the perturbation velocity $u_0 = 3.096$ m/s to $u_{max} = 120$ m/s from Table I. Physically, the interface has higher propensity to form resonance supersolid, which is inherently from the fact that there is geometric morphology at contact that easily satisfies the requirements on the spatial condition (54). As soon as there is a temporal wave passing through, those spatial morphologies shall form the waves in space and excites the interfaces.

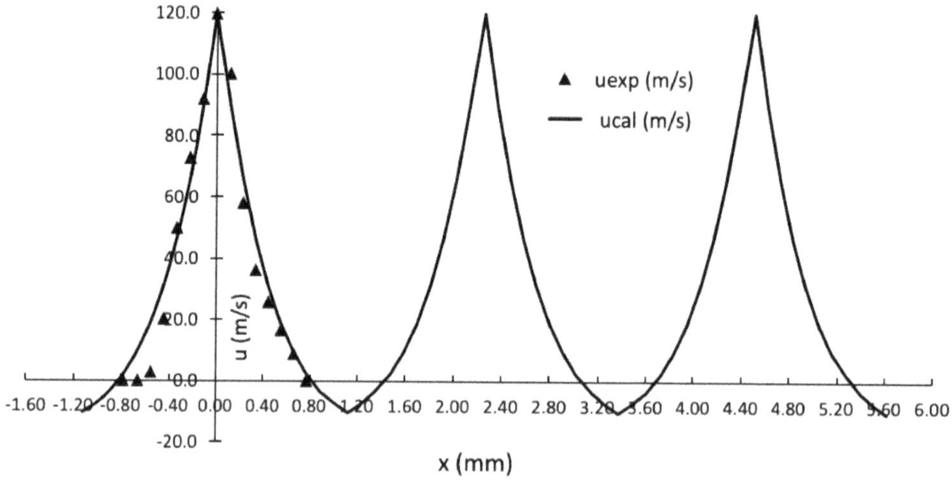


Figure 4: A pulse train at the interface.

## 6 CONCLUSIONS

In this work, we introduce the nondissipative dynamics in solids to solve the interface problem of in-plane velocities. The process, the governing differential equations and the solutions of nondissipative motions show that at interface, the nondissipative dynamics is a common phenomenon.

The purposes of this demonstration are two folds. First, the interfacial strain and deformation are mainly from the motion of nondissipative dynamics and can only be properly counted by the introduction of the nondissipative dynamics. This finding should greatly relax the demand on the boundary element method and the formulation of the nondissipative dynamics, sliding and topologies can be included in the integration equations of the boundary element method.

Second, the formulation of velocity and displacements at interfaces can be calculated by (83). The calculation of the nondissipative dynamics should be performed with the dissipative solid mechanics.

ACKNOWLEDGEMENTS

The authors are grateful for the support of the Beijing Natural Science Foundation IS23027.

REFERENCES

[1]
[2] Popov, V.L., *Contact Mechanics and Friction, Physical Principles and Applications*, Springer: Berlin, 2010.
[3] Burger, H., Forsbach, F. & Popov, V.L., Boundary element method for tangential contact of a coated elastic half-space. *Machines*, **11**(694), pp. 1–15, 2023.
[4] Komvopoulos, K., A multiscale theoretical analysis of the mechanical thermal, and electrical characteristics of rough contact interfaces demonstrating fractal behavior. *Frontiers in Mechanical Engineering*, **6**(36), pp. 1–20, 2020.
[5] Chen, J., Xu, X., Zhou, J. & Li, B., Interfacial thermal resistance: Past, present, and future. *Reviews of Modern Physics*, **94**, 025002, 2022.
[6] Stoner, R. & Maris, H., Kapitza conductance and heat flow between solids at temperatures from 50 to 300 K. *Physical Review B*, **48**(22), pp. 16373–16387, 1993.
[7] Swartz, E. & Pohl, R., Thermal boundary resistance. *Reviews of Modern Physics*, **61**, p. 605, 1989.
[8] Cahill, D.G. et al., Nanoscale thermal transport II: 2003–2012. *Applied Physics Reviews*, **1**, 011305, 2014.
[9] Ransom, J.B., McCleary, S.L. & Aminpour, M.A., A new finite element for connecting independently modeled substrates. *AIAA/ASME/ASCE/AHS/ASC Structures, Structural Dynamics/Material Conference*, 1993.
[10] D'Andrea, A., Tozzo, C., Boschetto, A. & Bottini, L., Interface roughness parameters and shear strength. *Modern Applied Science*, **7**(10), pp. 1–10, 2013.
[11] Kapitza, P., The study of heat transfer in helium II. *J. Phys. USSR*, **4**, p. 181, 1941.
[12] Landau, L., Theory of the superfluidity of helium II. *J. Phys. USSR*, **5**, pp. 301–330, 1941.
[13] Gol'danskii, V., Trakhtenberg, L. & Fleurov, V., *Tunneling Phenomena in Chemical Physics*, Taylor and Francis: New York, 2021.
[14] Su, S., Zhang, Y., Chen, J. & Shih, T.-M., Thermal electron-tunneling devices as coolers and amplifiers. *Nature Scientific Reports*, **6**, 21425, 2016.
[15] Bramble, J.H. & King, J.T., A finite element method for interface problems in domains with smooth boundaries and interfaces. *Advances in Computational Mathematics*, **6**, pp. 109–138, 1996.
[16] Aminpour, M.A., Krishnamurthy, T. & Shin, Y., Coupling of independent modeled three-dimensional finite element meshes with non-matching arbitrary shape interface boundaries. *AIAA Conference*, 1999.
[17] Fadale, T.D. & Aminpour, M.A., Thermal interface element for independently modeled finite element meshes. *AIAA Structures, Structural Dynamics/Material Conference*, 1999.
[18] Cheng, A.H.-D. & Cheng, D.T., Heritage and early history of the boundary element method. *Engineering Analysis with Boundary Elements*, **26**, pp. 268–302, 2005.
[19] Gipson, G.S., Issues related to interface calculations in the boundary element method. *Transactions on Modelling and Simulations*, 1999.
[20] Huang, D., Yan, X., Larsson, R. & Almqvist, A., Boundary element method for the elastic contact problem with hydrostatic load at the contact interface. *Applied Surface Science Advances*, **6**, 100176, 2021.

[21] Tan, C. & Gao, Y., Treatment of bimaterial interface crack problems using the boundary element method. *Engineering Fracture Mechanics*, **36**(6), pp. 919–932, 1990.
[22] Hutchings, I.M., Leonardo da Vinci's studies of friction. *Wear (Supplement C)*, **360**, pp. 51–66, 2016.
[23] Archard, J., Contact and rubbing of flat surfaces. *Journal of Applied Physics*, **24**(8), pp. 981–988, 1953.
[24] Earies, S. & Lee, C., Instabilities arising from the friction interaction of a pin-disk system resulting in noise generation. *Journal of Engineering for Industry*, **98**(1), pp. 81–86, 1976.
[25] Prakash, V., Frictional response of sliding interfaces subjected to time varying normal pressures. *Journal of Tribology*, **120**, pp. 97–102, 1998.
[26] Gupta, A. & Gupta, A.K., Missing mass effect in coupled analysis I: Complex modal properties. *Journal of Structural Engineering*, **124**(5), pp. 490–495, 1998.
[27] Oden, J. & Pires, E., Nonlocal and nonlinear friction laws and variational principles for contact problems in elasticity. *Journal of Applied Mechanics*, **50**, pp. 67–73, 1983.
[28] Chen, W., Hydrodynamic heat transfer in solids. *International Journal of Heat and Mass Transfer*, **215**, 124455, 2023.
[29] Malvern, L.E., *Introduction to the Mechanics of a Continuous Medium*, Prentice-Hall: Englewood, 1969.
[30] Bertram, A. & Gluge, R., *Solid Mechanics, Theory, Modeling and Problems*, Springer: Heidelberg, 2015.
[31] Chen, W., Heat transfer at speed of sound. *International Journal of Heat and Mass Transfer*, **177**, 121529, 2021.
[32] Chen, W., Heat transfer at speed of sound. *International Journal of Heat and Mass Transfer*, vol. **177**, pp. 1–13, 2021.
[33] Leggett, A., Can a solid be 'superfluid'? *Physical Review Letters*, **25**(22), pp. 1543–1546, 1970.
[34] Balibar, S. & Caupin, F., Supersolidity and disorder. *Journal of Physics on Condense Matters*, 20, 173201, 2008.
[35] Chen, W., On Taylor correlation functions in isotropic turbulent flows. *Nature Scientific Reports*, **13**(3859), pp. 1–31, 2023.
[36] Shimizu, T., Constants of the motion in nondissipative and dissipative systems. *Progress of Theoretical Physics*, **47**(4), pp. 1181–1199, 1972.
[37] Silva, R.P.d., Non-dissipative system as limit of a dissipative one: comparison of the asymptotic regimes. *Bulletin of the Brazilian Mathematical Society, New Series*, **51**, pp. 125–137, 2020.
[38] Chen, G.-Q.G., On degenerate partial differential equations. arXiv:1005.2713v1, **5**, pp. 1–39, 2010.
[39] Debussche, A., Hofmanova, M. & Vovelle, J., Degenerate parabolic stochastic partial differential equations: Quasilinear case. *The Annals of Probability*, **44**(3), pp. 1916–1955, 2016.
[40] Ivanov, A., On singularity points of equations of mechanics. *Doklady Mathematics*, **97**(2), pp. 167–169, 2018.
[41] Zhou, Y. & Gupta, V., Interface solutions of partial differential equations with point singularity. *Journal of Computational and Applied Mathematics*, **362**, pp. 400–409, 2019.

[42] Coker, D., Lykotrfitis, G., Needleman, A. & Rosakis, A., Frictional sliding modes along an interface between identical elastic plates subject to shear impact loading. *Journal of the Mechanics and Physics of Solids*, **53**(1), pp. 884–922, 2005.
[43] Lykotrafitis, G., Rosakis, A.J. & Ravichandran, G., Self-healing pulse-like shear rupture in the laboratory. *Science*, **313**, pp. 1765–1768, 2006.
[44] Rosakis, A., Samudrala, O. & Coker, D., Cracks faster than the shear wave speed. *Science*, **284**, pp. 1337–1340, 1999.
[45] Rosakis, A.J., Samudrala, O. & Coker, D., Intersonic shear crack growth along weak planes. *Materials Research Innovations*, **3**, pp. 236–243, 2000.

# FAST ALGORITHM FOR BOUNDARY INTEGRAL EQUATION SOLVING IN TWO-DIMENSIONAL FLOW SIMULATION BY VORTEX METHODS

ALEXANDRA KOLGANOVA & ILIA K. MARCHEVSKY
Bauman Moscow State Technical University, Russia

ABSTRACT
In the case of low subsonic velocities, when the compressibility of the medium can be neglected, vortex methods can be efficiently applied to simulate flows and estimate hydrodynamic loads acting on airfoils. The primary variable in vortex method is vorticity field, while the velocity and pressure fields can be reconstructed; there is also the possibility of computation of integral force and momenta acting on the airfoil, as well as viscous friction forces. The use of vortex methods requires multiple (at each time step) solution of the boundary integral equation (BIE) that describes the generation of vorticity on the airfoil surface, solution of several problems of $N$-body type and some other operations. Direct algorithms for these sub-problems have quadratic computational complexity with respect to number of particles, that significantly bounds the applicability of vortex methods for solving problems that require detailed discretisation. This paper presents description of fast algorithms based both on well-known approaches and some original modifications. The basic algorithm is hybrid algorithm for simulation of interaction of vortex particle in the flow domain; its generalisation is suggested for efficient solving of the BIE arising in vortex methods. The quasi-linear computational complexity of all suggested algorithms is achieved; their implementation is developed for multi-core CPUs.
*Keywords: vortex methods, fast algorithm, boundary integral equations, n-body problem.*

## 1 INTRODUCTION

Let us consider two-dimensional problems related to simulating of flow of a liquid or gas medium around airfoils and systems of airfoils. In the simplest case, the airfoils are rigid and immovable, but one can also consider the case with movable or deformable airfoils, including the fluid–structure interaction (FSI) problems, when the motion of the airfoil does not follow a predetermined law, but is determined by the hydrodynamic loads acting on it. Such problems often arise in engineering applications; the main purpose of analysis in them is to determine the unsteady loads acting on structural elements and characteristics of their oscillations (vibrations).

If the flow velocity is much lower than the velocity of sound, so the influence of the compressibility of the medium can be neglected, then meshless Lagrangian vortex methods [1]–[3] can be quite efficient, especially when simulating external flows. For a certain class of problems, vortex methods represent a powerful engineering tool, since their computational cost seems to be rather low in comparison to 'traditional' mesh-based numerical methods.

Direct implementation of the algorithms of vortex methods leads to quadratic computational complexity with respect to number of vortex particles. Two main sub-problems should be solved at each time step: vortex particles (i.e., existing vorticity) motion simulation and vorticity generation that takes place at airfoil boundaries and provides the no-slip boundary condition satisfaction. The governing equations that describe vortex particles motion in the flow are similar to the Newtonian's gravity low, so the considered problem is similar to the classical $N$-body problem. Well-known approximate fast methods (e.g., the Barnes–Hut method [4] and Fast multipole method [5]) of quasilinear computational complexity can be used, but specifically for 2D vortex method original hybrid algorithm has been developed [6].

WIT Transactions on Engineering Sciences, Vol 136, © 2024 WIT Press
www.witpress.com, ISSN 1743-3533 (on-line)
doi:10.2495/BE470091

The aim of this paper is to develop modification (generalisation) of the mentioned hybrid method for solving of the boundary integral equation that arises when simulating vorticity generation. In addition to quasilinear computational complexity, it is necessary to provide low requirements to memory usage since direct storage of the linear system matrix can require tens or hundreds of gigabytes of RAM.

## 2 MATHEMATICAL MODEL OF VORTEX METHODS

The mathematical model underlying the vortex methods includes the continuity equation in the form of incompressibility condition and the Navier–Stokes equations [7], which describe the motion of a viscous incompressible fluid:

$$\nabla \cdot \mathbf{V} = 0,$$

$$\frac{\partial \mathbf{V}}{\partial t} + (\mathbf{V} \cdot \nabla)\mathbf{V} = -\frac{\nabla p}{\rho} + \nu \Delta \mathbf{V}.$$

Here $\mathbf{V} = \mathbf{V}(\mathbf{r}, t)$ is the flow velocity field; $p = p(\mathbf{r}, t)$ is pressure field; $\rho = \text{const}$ is density; $\nu = \text{const}$ is kinematic viscosity. These equations should be supplemented with initial and boundary conditions: the perturbation decay condition on infinity and the no-slip condition on the walls.

The idea of vortex methods is to consider the vorticity field $\mathbf{\Omega} = \nabla \times \mathbf{V}$ as a primary computational variable, that allows in 2D case to reduce the problem to transfer simulation of one scalar field; the velocity and pressure fields can be reconstructed according to the generalised Helmholtz decomposition [8] and generalised Cauchy–Lagrange integral [9].

For viscosity effect simulation the Viscous Vortex Domains (VVD) method is used, that is based on the diffusive velocity approach [10]. Thus, the Navier–Stokes equations in two-dimensional case for incompressible flow can be written in terms of vorticity:

$$\frac{\partial \Omega}{\partial t} + \nabla \times \big(\mathbf{\Omega} \times (\mathbf{V} + \mathbf{W})\big) = \mathbf{0} \tag{1}$$

and interpreted as the equation for the transfer of vorticity $\mathbf{\Omega}$ along the field $(\mathbf{V} + \mathbf{W})$. The variable $\mathbf{W} = -\nu\frac{\nabla \Omega}{\Omega}$ is responsible for the diffusive velocity and is proportional to kinematic viscosity coefficient $\nu$; $\Omega$ is the component of vorticity vector $\mathbf{\Omega}$, that is orthogonal to the flow plane.

Such representation allows one to propose an efficient way to simulate the motion of vorticity in a viscous fluid: the vorticity is discretised into small-sized domains with total vorticity $\Gamma_i$. Domains which we call hereinafter 'vortex particles' are then moved along the velocity field $(\mathbf{V} + \mathbf{W})$, and the total vorticity remains unchanged. A discrete analogue of eqn (1) is the system of ordinary differential equations

$$\frac{d\mathbf{r}_i}{dt} = \mathbf{V}_i + \mathbf{W}_i, \quad \Gamma_i = \text{const}, \quad i = 1, \ldots, N.$$

Here $N$ is the number of vortex particles used to simulate the vorticity distribution.

### 2.1 Velocity field reconstruction

It is possible to reconstruct the velocity field using the generalised Helmholtz decomposition [8], which also can be considered as generalised Biot–Savart law. For unbounded flow domain and vorticity in form of small-sized vortex particles, one can use the following formula for velocity reconstruction:

$$\mathbf{V}(\mathbf{r}) = \mathbf{V}_\infty + \sum_{j=1}^{N} \left(\mathbf{k} \times \mathbf{Q}_\varepsilon(\mathbf{r} - \mathbf{r}_j)\right) \Gamma_j = \mathbf{V}_\infty + \sum_{j=1}^{N} \frac{\mathbf{k} \times (\mathbf{r} - \mathbf{r}_j)}{2\pi \max\left\{|\mathbf{r} - \mathbf{r}_j|^2, \varepsilon^2\right\}} \Gamma_j.$$

Here $\mathbf{V}_\infty$ is the incident flow velocity; $\mathbf{r}_j$ is position of the *j*th vortex particle; $\mathbf{k}$ is unit normal vector to the flow plane; $\mathbf{Q}_\varepsilon$ is regularised gradient of the fundamental solution of the Laplace equation (sometimes also called 'the Biot–Savart law kernel')

$$\mathbf{Q}(\mathbf{r} - \mathbf{r}_j) = \frac{\mathbf{r} - \mathbf{r}_j}{2\pi |\mathbf{r} - \mathbf{r}_j|^2};$$

$\varepsilon$ is small vortex particle radius.

The calculation of the velocities $\mathbf{V}(\mathbf{r}_i)$ of all the vortex particles is similar to solution of the *N*-body problem, as it requires the computation of their mutual interactions. In terms of the number of operations, this part of the algorithm is one of the most time-consuming, since the direct algorithm has quadratic computational complexity $O(N^2)$. To speed up this operation, approximate fast algorithms with quasilinear computational complexity are used $O(N \log^\alpha N)$ [4]–[6].

### 2.2 Vorticity generation

The next most important and time-consuming operation in vortex methods is the solution at each time step of the boundary integral equation that describes the generation of vorticity at the airfoil boundary.

The vorticity generated at each time step of simulation is considered as a thin vortex sheet placed on the airfoil boundary (to be more precise, it is placed in the flow domain on infinitesimal distance from the airfoil). The intensity of the vortex sheet is chosen in such a way to satisfy the no-slip condition for the velocity field. Thus, one can derive the vector boundary integral equation [8]

$$\oint_K \left(\boldsymbol{\gamma}(\boldsymbol{\xi}) \times \boldsymbol{Q}(\mathbf{r} - \boldsymbol{\xi})\right) dS_\xi - \alpha(\mathbf{r})\left(\boldsymbol{\gamma}(\mathbf{r}) \times \boldsymbol{n}(\mathbf{r})\right) = \boldsymbol{f}(\mathbf{r}), \quad \mathbf{r} \in K. \tag{2}$$

Here $\boldsymbol{\gamma}(\mathbf{r})$ is unknown vortex sheet intensity distribution; $\alpha(\mathbf{r})$ is a coefficient responsible to the smoothness of the boundary that follows from the Sokhotski–Plemelj theorem [8]. The right-hand side function $\boldsymbol{f}(\mathbf{r})$ depends both on the velocity of the airfoil boundary and influences of all the vortex particles in the flow. Its calculation at number of points on the airfoil boundary is similar to influence calculation of *N* bodies at some specified points.

Let us project eqn (2) onto tangent direction at each point of the boundary, that is equivalent to scalar multiplication by the vector $\boldsymbol{\tau}(\mathbf{r})$. The resulting scalar integral equation, which is called the *T*-model [11] has the following form:

$$\oint_K \left( \underbrace{\left(\boldsymbol{k} \times \boldsymbol{Q}(\mathbf{r} - \boldsymbol{\xi})\right) \cdot \boldsymbol{\tau}(\mathbf{r})}_{P_\tau(\mathbf{r},\boldsymbol{\xi})} \right) \gamma(\boldsymbol{\xi})\, dl - \alpha(\mathbf{r})\gamma(\mathbf{r}) = \underbrace{\boldsymbol{f}(\mathbf{r}) \cdot \boldsymbol{\tau}(\mathbf{r})}_{f_\tau(\mathbf{r})}, \quad \mathbf{r} \in K \tag{3}$$

To solve the boundary integral equation, eqn (3), a family of numerical schemes (*T*-schemes) was developed based on the Galerkin method [11], which are based on the fact that the kernel of the integral eqn (3) is bounded or has a weak (absolutely integrable) singularity, and the airfoil is replaced with polygon with rectilinear segments, called hereinafter 'panels'.

As a result of usage of the $T$-schemes, it is possible to construct a system of linear algebraic equations that is a discrete analogue of the integral equation, eqn (3).

The right-hand side of the linear system represents the averaged over panels velocity influenced by all the vortex particles in the flow domain (and includes some other integral terms in the case of movable airfoil). The computational complexity of the right-hand side calculation is proportional to the product of number of panels $M$ and number of vortex particles $N$. Again, we obtain a problem of quadratic complexity $O(MN)$, similar to the calculation of mutual interactions of particles.

The calculation of the right-hand side becomes especially time-consuming when the number of vortex particles increases and more detailed airfoil discretisation is used. The solution of such a problem by direct method turns out to be inefficient. Therefore, it makes sense to use fast algorithms to calculate the right-hand side.

It is suggested to use iterative methods to solve the obtained linear system: the solution procedure requires at each iteration matrix-to-vector multiplication, which direct implementation is also time-consuming operation since the matrix is dense. At the same time such matrix-to-vector multiplication can be considered as a procedure similar to calculating the effect of vortex particles on each other. Consequently, when solving systems of linear equations, matrix-to-vector multiplication can also be performed by a fast method.

In the considered case let us use the generalised method of minimum residuals, GMRES [12] as an iterative method. It is applicable to systems with arbitrary matrices and there is no need for multiplication of a transposed matrix, which makes it possible to apply fast algorithm.

## 3 FAST ALGORITHMS

There are various fast algorithms and approaches that allow to solve $N$-body type problems. It seems that the most suitable are fast methods based on hierarchical partitioning of the domain with constructing of a tree, which traversal allows for approximate calculating of the necessary quantities. From this class of 'Treecodes' fast methods let us highlight the Barnes–Hut method [4] and the fast multipole method [5].

The main idea of the classical Barnes–Hut method is constructing of a $k$-d or quad-tree and its traversal upward: from the leaves to the root, during which all cell characteristics are calculated: for lower-level cells by processing all particles included in them, for higher-level cells – by processing the characteristics of child cells. Then, for each particle, the tree is traversed downward: from root to leaves. Far located particles are combined into clusters, which influence is calculated approximately as from one 'summary' particle; closely located particles are processed directly: the influence from each particle is calculated straightforwardly. The computational complexity of the algorithm is $O(N \log N)$.

The fast multipole method is based on similar ideas, but all cells at the same level of the tree hierarchy have strictly the same size that is achieved by constructing a quadtree. The upward traversal of the tree is similar to the BH method with the only difference that not only total mass/charge is calculated, but also number of the so-called multipole moments. The downward traversal is performed once, and in each cell a local expansion of the influence of all sufficiently distant cells is constructed. For the lower-level cells (leaves), the influence of particles from neighbouring cells is calculated directly. The computational complexity of the algorithm is $O(N)$.

In this paper let us consider an original author's modification in the form of a hybrid BH/FMM method [6], which is similar to the BH method in the sense of tree traversal logic, but also uses some ideas of the FMM method:

- representation of the particles in cells with 'long' multipole expansions instead of the influence of a single particle;
- formation of local expansions in leaf-cells, which eliminates the need of the tree traversing for each particle; the downward traversal is performed only for the leaves of the tree.

The flow domain is divided into cells, that corresponds to the construction of a $k$-d tree in the computational domain. The tree is traversed firstly upward, and then downward for each leave. If the influencing cell (cluster) is far enough from the control cluster, its summary influence can be approximated with high accuracy with the multipole expansion of the velocity function – with respect to negative powers of the distance $|\boldsymbol{\rho}|$ between cluster centres (Fig. 1).

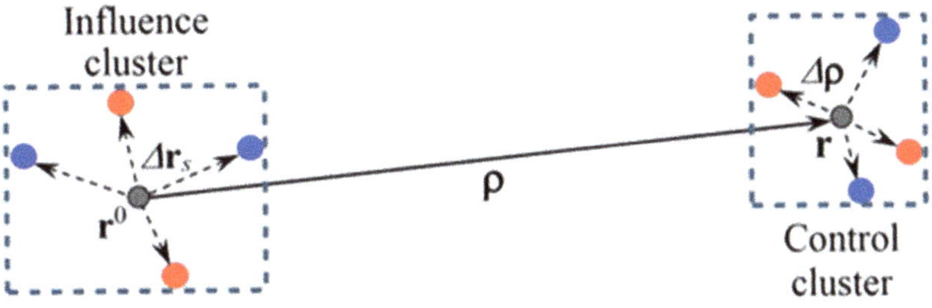


Figure 1: Influencing and control clusters.

The velocity function is expanded into power series using the Taylor's formula, considering that the distance $|\boldsymbol{\rho}|$ between the control and influence clusters is much larger than the distances between vortices within the influence cluster:

$$\mathbf{V}(\mathbf{r}) = \mathbf{V}^{\mathrm{m}}(\boldsymbol{\rho}) + \mathbf{V}^{\mathrm{d}}(\boldsymbol{\rho}) + \mathbf{V}^{\mathrm{q}}(\boldsymbol{\rho}) + \mathbf{V}^{\mathrm{o}}(\boldsymbol{\rho}) + \mathbf{V}^{\mathrm{h}}(\boldsymbol{\rho}) + \cdots.$$

Here the terms $\mathbf{V}^{\mathrm{m}}$, $\mathbf{V}^{\mathrm{d}}$, $\mathbf{V}^{\mathrm{q}}$, $\mathbf{V}^{\mathrm{o}}$, $\mathbf{V}^{\mathrm{h}}$ are called monopole, dipole, quadrupole, octupole and hexadecapole, etc., and their values (in 2D case) can be estimated in order of magnitude as

$$|\mathbf{V}^{p}| \sim \left(\frac{|\Delta\mathbf{r}|}{|\boldsymbol{\rho}|}\right)^{p+1},$$

where $|\Delta\mathbf{r}|$ is the typical size of influencing cluster; index $p$ corresponds to considered multipole terms: $p = 0$ for monopole term, $p = 1$ for dipole, $p = 2$ for quadrupole, etc.

When simulating two-dimensional flows, the position of the vortex particle $\mathbf{r}_s = \{x_s, y_s\}$, $s = 1, \dots, N$, can be considered as a complex number $r_s = x_s + i\,y_s$. Let us define the concept of multipole moments using complex numbers. The monopole moment of a tree cell (cluster) is a scalar real value $m^{(0)}$, while all higher-order moments can be represented through complex numbers $m^{(p)} = m_0^{(p)} + i\,m_1^{(p)}$, where $p \geq 1$. Thus, all multipole moments of leaf cells with respect to their centres are calculated through vortex particle circulations $\Gamma_s$ and complex numbers $\Delta r_s$, that correspond to their positions $\Delta\mathbf{r}_s$, $s = 1, \dots, N$, as follows:

$$m^{(p)} = \textstyle\sum_{s=1}^{N} \Gamma_s \Delta r_s^{\,p}, \quad p \geq 0, \tag{4}$$

complex numbers $\Delta r_s$ correspond to vectors $\Delta\mathbf{r}_s$ (Fig. 1).

For parent cells multipole moments are calculated by summarising the corresponding moments of child cells, which should be shifted to the centre of the parent cell, similar to fast multipole method:

$$m_h^{(p)} = \sum_{k=0}^{p} C_p^k m^{(p-k)} h^k, \quad p \geq 0 \tag{5}$$

where $C_p^k$ is binomial coefficient; $h$ is complex number that corresponds to shifting vector $\mathbf{h}$ (Fig. 2). It is sufficient to calculate all multipole terms in the velocity expansion once for each cluster, and then use them many times to approximate calculation of the influence of this cluster at arbitrary observation points.

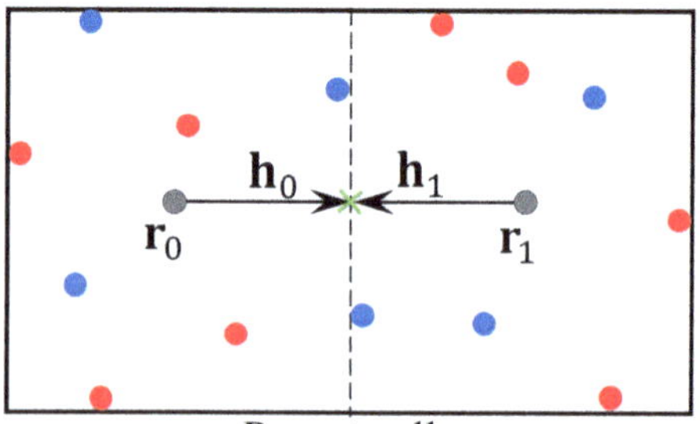


Figure 2: Multipole moments shifting to the centre of the parent cell.

To approximately calculate the velocities of particles in a control cluster, it is enough to construct a local expansion in the vicinity of its centre (Fig. 1):

$$\mathbf{V}(\mathbf{r} + \Delta\boldsymbol{\rho}) \approx \frac{\mathbf{k}}{2\pi} \times \sum_{k=0}^{p} \mathbf{U}^{(k)}, \tag{6}$$

where $\mathbf{U}^{(k)}$ is a vector that corresponds to complex number $U^{(k)} = \frac{E_k}{k!}(\Delta\rho^*)^k$,

$$E_k = (-1)^k \sum_{q=0}^{K} \frac{\Theta^{(k+q)}}{q!} \left(m^{(q)}\right)^*, \tag{7}$$

complex coefficients $\Theta^{(q)}$ are calculated recursively,

$$\Theta^{(0)} = \frac{\rho}{|\boldsymbol{\rho}|^2}, \quad \Theta^{(q)} = \frac{q}{|\boldsymbol{\rho}|^2} \Theta^{(q-1)} \rho, \quad q = 1, 2, \dots, \tag{8}$$

$\rho$ is complex number corresponding to vector $\boldsymbol{\rho}$; the asterisk '*' means complex conjugation.

If the influencing cluster is located close to the control one, its influence at all observation points is calculated directly according to the Biot–Savart law.

The proximity criterion is given by the condition $|\boldsymbol{\rho}| \leq \frac{h_p + h_q + \varepsilon}{\theta}$, where $\rho$ is the distance between the geometric centres of the control and influence cells; $h_p$ and $h_q$ are sums of width and length of the corresponding cells; vortices radius $\varepsilon$ is added to numerator for the correct processing of cells containing a single particle, since the values of $h_p$ and $h_q$ for such cells are equal to zero. The proximity parameter $\theta$ determines the relationship between accuracy and computational complexity: the smaller the value of $\theta$ is chosen, the smaller the error the fast algorithm provides by increasing the size of the close zone.

For the *N*-body problem, when number of bodies is less than $10^6$, the suggested hybrid BH/FMM method shows higher performance in comparison to the FMM method (Fig. 3).

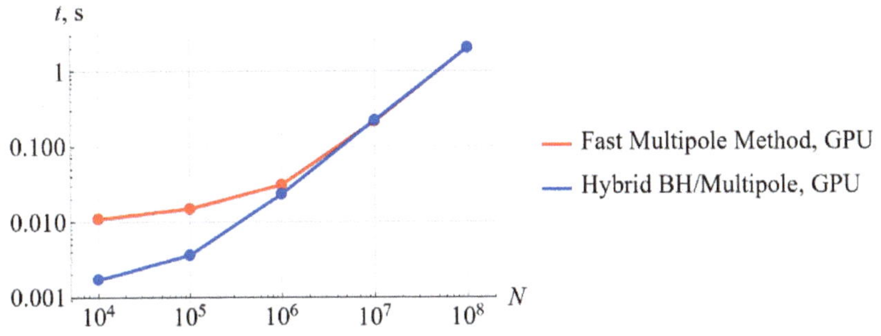


Figure 3: Computational time in *N*-body problem.

## 4 BOUNDARY INTEGRAL EQUATION SOLVING

As discussed earlier, the solution of the boundary integral equation is reduced to a system of linear algebraic equations thanks to a family of numerical algorithms called *T*-schemes [11]. Let us consider piecewise-linear numerical scheme $T^1$.

The numerical solution on the *i*th panel has the following form.

$$\gamma(\mathbf{r}) = \frac{\Gamma_i^0}{L_i}\varphi_i^0(\mathbf{r}) + \frac{\Gamma_i^1}{L_i}\varphi_i^1(\mathbf{r}), \quad r \in K_i,$$

function $\varphi_i^0(\mathbf{r})$ is equal to 1 over the *i*th panel, $\varphi_i^1(\mathbf{r})$ varies linearly from $(-1/2)$ to $1/2$ along the *i*th panel; $L_i$ is length of the *i*th panel. According to the Galerkin method, one should orthogonalise the residual of eqn (3) to both families of functions, i.e., $\{\varphi_i^0(\mathbf{r})\}_{i=1}^M$ and $\{\varphi_i^1(\mathbf{r})\}_{i=1}^M$, where $M$ is total number of panels.

The resulting linear system has the following block structure:

$$\begin{pmatrix} [A^{00}] + [D^{00}] & [A^{01}] & \{I\} \\ [A^{10}] & [A^{11}] + [D^{11}] & \{O\} \\ \{I\}^T & \{O\}^T & 0 \end{pmatrix} \begin{pmatrix} \{\Gamma^0\} \\ \{\Gamma^1\} \\ R \end{pmatrix} = \begin{pmatrix} \{b^0\} \\ \{b^1\} \\ \Gamma \end{pmatrix} \quad (9)$$

where $R$ is regularising variable [3]; $\{I\}$ is the column consists of unities; $\{O\}$ is zero column; each matrix coefficient $a_{ij}^{\alpha\beta}$ of the corresponding block $[A^{\alpha\beta}]$ expresses influence of the *j*th (influencing) panel with a constant ($\beta = 0$) or linear ($\beta = 1$) vorticity distribution with unit intensity onto the *i*th (control) panel with constant ($\alpha = 0$) or linear ($\alpha = 1$) weight (projection) function. To calculate them, one should integrate the influence function

$$\mathbf{Q}(\mathbf{r} - \boldsymbol{\xi}) = \frac{\mathbf{r} - \boldsymbol{\xi}}{2\pi|\mathbf{r} - \boldsymbol{\xi}|^2}$$

twice – over the influencing and control panels with the necessary weights:

$$a_{ij}^{\alpha\beta} = \frac{\mathbf{n}_i}{L_j} \cdot \int_{K_i} \left( \int_{K_j} \mathbf{Q}(\mathbf{r} - \boldsymbol{\xi})\, \varphi_j^\beta(\boldsymbol{\xi}) dl_\xi \right) \varphi_i^\alpha(\mathbf{r}) dl_r \quad (10)$$

where indices $\alpha$ and $\beta$ corresponds to constant and linear basis and projection functions, respectively; the right-hand side coefficients are expressed through integrals over panels:

$$b_i^\alpha = \boldsymbol{\tau}_i \cdot \left( \sum_{s=1}^N \Gamma_s \int_{K_i} \mathbf{Q}(\mathbf{r} - \mathbf{r}_s) \varphi_i^\alpha(\mathbf{r}) dl_r \right) \quad (11)$$

where $\mathbf{n}_i$ and $\boldsymbol{\tau}_i$ are outer normal and tangent (counterclockwise direction) unit normal vectors for the *i*th panel; $\Gamma_s$ and $\mathbf{r}_s$ are circulation and position of the vortex particle in the flow. Blocks $[D^{00}]$ and $[D^{11}]$ are diagonal matrices consist of the following values:

$$d_{ii}^{00} = -\frac{1}{2L_i}, \quad d_{ii}^{11} = -\frac{1}{24L_i}.$$

4.1 The right-hand side computation

If there was no integration in brackets in eqn (11), the calculation of the sum would be equal exactly to calculation of the influence of *N* particles at specified point $\mathbf{r}$. This could be done by using the discussed fast method according to eqn (4). To provide integration, we integrate analytically the local expansion, calculated in cluster that involves the *i*th panel. Eqns (6)–(8) remain correct, but now eqn (6) becomes responsible for integrated velocity over panel, and one should replace $U^{(k)}$ for constant projection function $\varphi_i^0(\mathbf{r})$ with

$$U^{(k,0)} = \sum_{s=0}^{p} \frac{E_S}{(s+1)!}\frac{\tau_i}{L_i}\left(\left(\Delta\rho_{i,\mathrm{end}}^*\right)^S - \left(\Delta\rho_{i,\mathrm{beg}}^*\right)^S\right) \tag{12}$$

and for linear projection function $\varphi_i^1(\mathbf{r})$ with

$$U^{(k,1)} = \sum_{s=0}^{p} \frac{E_S}{(s+2)!}\left(\frac{\tau_i}{L_i}\right)^2 \left(\left(\Delta\rho_{i,\mathrm{end}}\right)^s \left(\frac{(s+1)\tau_i L_i}{2} - \Delta\rho_i\right) - \right.$$
$$\left. - \left(\Delta\rho_{i,\mathrm{beg}}\right)^S \left(\frac{(s+1)\tau_i L_i}{2} + \Delta\rho_i\right)\right)^* \tag{13}$$

where $\tau_i$ is complex number corresponding to the vector $\boldsymbol{\tau}_i$; $\Delta\rho_i$, $\Delta\rho_{i,\mathrm{beg}}$ and $\Delta\rho_{i,\mathrm{end}}$ are complex numbers corresponding to the vectors connecting the centre of the control cluster with the centre, beginning and ending of the *i*th panel, respectively.

4.2 Matrix-to-vector multiplication

Matrix coefficients $a_{ij}^{\alpha\beta}$ will not be calculated explicitly. Instead of it, matrix-to-vector multiplication

$$\begin{pmatrix} [A^{00}] & [A^{01}] \\ [A^{10}] & [A^{11}] \end{pmatrix} \begin{pmatrix} \{\Gamma^0\} \\ \{\Gamma^1\} \end{pmatrix} \tag{14}$$

will be performed implicitly.

Since iterative method is used for linear system solving, $\{\Gamma^0\}$ and $\{\Gamma^1\}$ are known from the previous iteration; their initial values are usually zeros. Each component of the resulting column in eqn (14) can be written down as

$$\sum_{j=1}^{M}\left(a_{ij}^{\alpha 0}\Gamma_j^0 + a_{ij}^{\alpha 1}\Gamma_j^1\right)$$
$$= \mathbf{n}_i \cdot \sum_{j=1}^{M} \int_{K_i}\left(\int_{K_j} \mathbf{Q}(\mathbf{r}-\boldsymbol{\xi})\left(\varphi_j^0(\boldsymbol{\xi})\frac{\Gamma_j^0}{L_j} + \varphi_j^1(\boldsymbol{\xi})\frac{\Gamma_j^1}{L_j}\right) dl_\xi\right)\varphi_i^\alpha(\mathbf{r})dl_r \tag{15}$$

The inner integral expresses the influence that is induced by panel with linear distribution of vortex sheet intensity; at far enough distance it can be calculated through multipole expansion. Thus, we obtain a formula similar to eqn (6):

$$\int_{K_j} \mathbf{Q}(\mathbf{r}-\boldsymbol{\xi})\left(\varphi_j^0(\boldsymbol{\xi})\frac{\Gamma_j^0}{L_j}+\varphi_j^1(\boldsymbol{\xi})\frac{\Gamma_j^1}{L_j}\right)dl_\xi \approx \frac{\mathbf{k}}{2\pi}\times\sum_{k=0}^{p}\mathbf{U}^{(k)}$$

Eqns (7) and (8) remain correct, the only difference is the rule of calculation of multipole moments of leaf cells: firstly, it is required to calculate multipole moments of all the panels in the cell with respect to the panels' centres; for the *j*th panel

$$m_j^{(k)} = \frac{1}{(k+1)}\left(\frac{\tau_j L_j}{2}\right)^k \Gamma_j^0,\ k = 0, 2, 4, \ldots \tag{16}$$

$$m_j^{(k)} = \frac{1}{2(k+2)}\left(\frac{\tau_j L_j}{2}\right)^k \Gamma_j^1,\ k = 1, 3, 5, \ldots \tag{17}$$

Then multipole moments of panels should be shifted to the centre of the leaf cell according to eqn (5) and summarised. Further calculation of moments for parent cells is identical to the case of vortex particles. Note, that we assume the panel to be included into the specific cell only if the centre of the panel is inside this cell.

The outer integral in eqn (15) is calculated by integrating the local expansion, so eqn (6) now becomes responsible for the value of the repeated integral, eqns (7) and (8) remain true, except of exchanging $U^{(k)}$ with eqns (12) and (13).

After calculating the result of repeated integration, it should be multiplied by $\mathbf{n}_i$. According to the structure of the left-hand side of linear system (9), additionally products $[D^{00}]\,\{\Gamma^0\}$ and $[D^{11}]\,\{\Gamma^1\}$ required to be calculated. The latter operation has extremely low computational cost since $[D^{00}]$ and $[D^{11}]$ are diagonal blocks. Multiplication of the last row of the matrix in eqn (9) by the column and the last column of the matrix by scalar variable *R* are also very 'cheap' operations.

Note, that the right-hand side should be calculated just once since it remains the same for all iterations, while matrix-to-vector multiplication (14) is performed at each iteration. So it is required to recalculate multipole moments for all cells of the tree, constructed for panels. Since the structure of the tree is not changed, such recalculations are not a bottleneck.

## 5 ALGORITHM FOR FAST SOLVING OF BOUNDARY INTEGRAL EQUATION

Let us point out the main steps of the fast algorithm of the BIE (3) solving.

1. Determination of two bounding boxes: the first contains all the particles in the flow domains, the second contains centres of all the panels on the airfoil(s). These boxes are root cells for two trees construction.
2. Construction of *k*-d trees. The most suitable way to do it is to calculate the Morton codes of the particles/panel centres and then treat their positions on fractal Morton's curve in consistent way [13].
3. Calculation of multipole moments of leaf cells of both trees. For particle tree: according to eqn (4). For panel tree: first, multipole moments of all the panels should be calculated according to eqns (16) and (17), then moments of leaf cells – by shifting according to eqn (5) and summarising panels' moments.
4. Upward tree traversal (for both trees) to compute the multipole moments of all parent cells by summarising moments of child cells, being shifted according to eqn (5).

5. Before iterations: downward particle tree traversal for each leaf cell of panel tree to calculate the right-hand side of linear system (9), as it described in Section 4.
6. Iterations until convergence.
   a. Downward panel tree traverse for all its leaf cells to calculate the result of matrix-to-vector multiplication (14).
   b. Solution $\{\Gamma^0\}$ and $\{\Gamma^1\}$ updating according to the used iterative method (note, that in GMRES solution is not updated explicitly at each iteration, instead of it the matrix is multiplied by some vector, that can be formally treated as iterative solution).
   c. Recalculating multipole moments of the panels by using updated values of $\{\Gamma^0\}$ and $\{\Gamma^1\}$.
   d. Recalculating multipole moments of all the cells of the panel tree.

If the airfoil is movable, in the right-hand side of eqn (3) additional terms arise, that can be treated as influence of (the so-called 'attached') vortex sheets and source sheets, placed on panels. So, these terms can be calculated by downward traversing the panel tree with introducing some additional properties to their cells.

Note that for particle tree small value of $\varepsilon$ in proximity criterion is normally chosen as typical distance between particles in flow domain, while for panel tree it seems to be suitable to perform the following trick: assume $\varepsilon = 0$ with extending sizes of cells in order to bound all the panels in them. Resulting tree contains overlapping cells, but it is not an issue for Barnes–Hut-type method.

## 6 NUMERICAL EXPERIMENT

Let us present the results of solving model problems for the airfoil being split into $M = 3{,}200$ and $M = 12{,}800$ panels (Table 1). The linear system, that corresponds to $T^0$ and $T^1$ schemes of the first and second order of accuracy, respectively, was solved using:

- the Gaussian elimination method (implemented in Intel MKL library),
- iterative GMRES method with direct matrix calculation and storing (in-house implementation with matrix-to-vector multiplication with Intel MKL library),
- iterative GMRES method with fast matrix-to-vector multiplication.

The 'in-house' implementation of GMRES corresponds exactly to [12] with Hessenberg-type matrix and Givens rotations algorithm at each iteration (without restarts).

Note, that it is possible to decrease number of iterations in GMRES by applying pre-conditioner matrix which consists of tri-diagonal parts of blocks $[A^{00}] + [D^{00}]$ and $[A^{11}] + [D^{11}]$ (also with their north-east and south-west corner elements) and the last line and the last column of the matrix of the system (9). For the $T^0$ scheme the preconditioner is constructed similarly. Portraits of the preconditioner matrices are shown in Fig. 4. To solve a system with such preconditioner modification of sweeping algorithms [14] is developed.

The calculations were performed on 18-cores Intel i9-10980XE processor in sequential and parallel regimes (OpenMP + embedded to Intel MKL parallelisation), in scheme $T^0$ the number of unknowns is equal to number of panels $M$, in scheme $T^1$ number of unknowns is twice the number of panels. Note that for $M = 3{,}200$ the errors of both schemes (including the error that follows from fast method usage) are $\delta^0 \sim 1.8 \cdot 10^{-4}$, $\delta^1 \sim 1.6 \cdot 10^{-6}$, for $M = 12{,}800$ one obtains $\delta^0 \sim 4.5 \cdot 10^{-5}$, $\delta^1 \sim 1.0 \cdot 10^{-7}$; $\delta^0$ and $\delta^1$ represent the relative errors of the fast method with respect to the result of the solution by the direct algorithm:

Table 1: Solution time in seconds and speedup in parallel multicore regime.

| Method | $M = 3{,}200$ | | $M = 12{,}800$ | |
|---|---|---|---|---|
| | Sequential (1 core) | Parallel (18 cores) | Sequential (1 core) | Parallel (18 cores) |
| | Scheme $T^0$ | | | |
| Gaussian elimination | 1.71 | 0.152 (×11.2) | 36.7 | 3.46 (×10.6) |
| GMRES + direct multiplication | 1.33 | 0.098 (×13.6) | 21.4 | 1.57 (×13.6) |
| GMRES + fast multiplication | 0.068 | 0.016 (×4.2) | 0.510 | 0.141 (×3.6) |
| | Scheme $T^1$ | | | |
| Gaussian elimination | 5.95 | 0.590 (×10.1) | 178 | 18.1 (×9.8) |
| GMRES + direct multiplication | 3.42 | 0.254 (×13.5) | 54.6 | 4.03 (×13.5) |
| GMRES + fast multiplication | 0.166 | 0.063 (×2.6) | 1.31 | 0.249 (×5.3) |

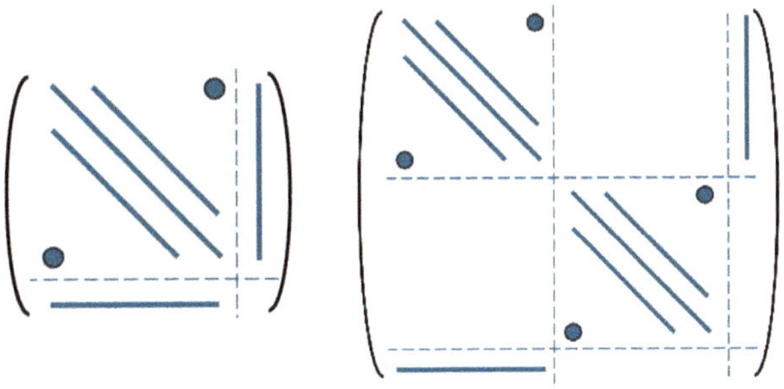

Figure 4: The preconditioner portraits for schemes $T^0$ and $T^1$.

$$\delta = \frac{\max_i |\gamma_i(\mathbf{r}) - \gamma_i^*(\mathbf{r})|}{\max_i |\gamma_i^*(\mathbf{r})|},$$

the indices '0' and '1' corresponds to piecewise-constant and piecewise-linear solutions.

The table shows that the speed up of calculations due to usage of the fast algorithm has order of tens, up to 100 times in comparison to GMRES implementation with direct matrix-to-vector multiplication for the considered airfoil discretisation. In turn, Gaussian elimination is additionally 1.5…3 times more time-consuming. Note, that for $T^0$ scheme speed up for fast multiplication algorithm is nearly the same for small and large problems (the value 4.2 in Table 1 is the averaged value, it varies due to small size of the matrix), while for $T^1$ scheme efficiency of parallelisation increases with matrix size growth.

A significant advantage of using a fast method for solving of the boundary integral equation (to be more precise, linear system which it corresponds to) is that there is no need to compute and store all the components of the matrix. It is enough to compute explicitly and store about only ~1% of the components (which are calculated directly, since the panels are close to each other), which makes the algorithm efficient not only in terms of computation time, but also in terms of the amount of memory consumed. For example, for $M = 3{,}200$ matrix storage requires 80 Mb of RAM for scheme $T^0$ and 320 Mb for scheme $T^1$, at the same time for $M = 12{,}800$ it requires 1.3 Mb and 5.2 Gb of RAM, respectively. If system of airfoils is considered, it can be impossible to store full matrix in memory.

The fast algorithm was generalised to simulate the flow around systems of airfoils. Table 2 shows a comparison of times of computation for a single airfoil split into 1 000 panels and a system of several identical airfoils using the GMRES method with direct and fast matrix multiplication for $T^1$-scheme with a piecewise-linear solution representation. Interesting to note, that in this problem speedup due to usage of the fast algorithm is nearly the same independently on total number of panels ($M$ = 1 000...9 000). Matrix-to-vector multiplication, however, has quasilinear complexity $O(M \log M)$, but number of iterations required for convergence of GMRES method increases from 3 to 13.

Table 2: Computation time for airfoil systems (in seconds) and number of iterations.

| | 1 airfoil | | 2 airfoils | | 4 airfoils | | 9 airfoils | |
|---|---|---|---|---|---|---|---|---|
| GMRES | Time | Iter. | Time | Iter. | Time | Iter. | Time | Iter. |
| Direct | 0.169 | 3 | 0.804 | 7 | 3.95 | 10 | 22.6 | 13 |
| Fast, $10^{-5}$ | 0.003 | 3 | 0.017 | 7 | 0.049 | 10 | 0.297 | 13 |
| Fast, $10^{-7}$ | 0.004 | 3 | 0.021 | 7 | 0.063 | 10 | 0.343 | 13 |

Note that large number of numerical experiments, having been performed for different types of problems solved with vortex particle methods [15] show that the minimal acceptable accuracy of the BIE solution has order of $10^{-5}$; for the problems with higher spatial and time resolution it should be $10^{-7}$. Higher accuracy is not required.

## 7 CONCLUSION

The fast algorithm is developed for numerical solution of the boundary integral equation that arises in two-dimensional problems of viscous incompressible flow simulation around airfoils. This algorithm is based on Galerkin-type numerical schemes for discretisation of the BIE ($T$-schemes, [11]) and can be considered as generalisation of the hybrid Barnes–Hut/ multipole fast algorithm [6] initially developed for simulation of interaction of vortex particles. For all time-consuming operations fast algorithms are developed: the right-hand side computation and matrix-to-vector multiplication, taking into account the specific features of the considered schemes $T^0$ and $T^1$. All other algorithms (tree construction, its traversals, etc.) are adapted for the BIE solution. The fast algorithm provides matrix-to-vector multiplication in GMRES method; efficient preconditioners are proposed. Parallel versions for all developed algorithms are developed for computations on multicore CPUs. The algorithms are flexible and can provide the required ratio of accuracy and complexity. The developed approaches allow one to achieve speedup of order of tens up to hundreds of times in practical applications with rather fine discretisation providing high accuracy.

## REFERENCES

[1] Cottet, G.H. & Koumoutsakos, P.D., *Vortex Methods: Theory and Practice*, 2000.
[2] Mimeau, C. & Mortazavi, I., A review of vortex methods and their applications: From creation to recent advances. *Fluids*, **6**(2), p. 68, 2021.
[3] Lifanov, I.K., *Singular Integral Equations and Discrete Vortices*, VSP: Utrecht, 1996.

[4] Barnes, J. & Hut, P., A hierarchical $O(n \log n)$ force-calculation algorithm. *Nature*, **324**(4), pp. 446–449, 1986.

[5] Greengard, L. & Rokhlin, V., A fast algorithm for particle simulations. *Journal of Computational Physics*, **73**(2), pp. 325–348, 1987.

[6] Marchevsky, I., Ryatina, E. & Kolganova, A., Fast Barnes–Hut-based algorithm in 2D vortex method of computational hydrodynamics. *Computers and Fluids*, **266**(1), 106018, 2023.

[7] Batchelor, G.K., *An Introduction to Fluid Dynamics*, Cambridge University Press, 1967.

[8] Kempka, S.N., Glass, M.W., Peery, J.S., Strickland, J.H. & Ingber M.S., Accuracy considerations for implementing velocity boundary conditions in vorticity formulations. *SANDIA Report*, SAND96-0583, 52 pp., 1996.

[9] Dynnikova, G.Y., An analog of the Bernoulli and Cauchy–Lagrange integrals for a time-dependent vortex flow of an incompressible fluid. *Fluid Dynamics*, **35**, pp. 24–32, 2000.

[10] Guvernyuk, S.V. & Dynnikova, G.Y., Modeling the flow past an oscillating airfoil by the method of viscous vortex domains. *Fluid Dynamics*, **42**(1), pp. 1–11, 2007.

[11] Marchevsky, I.K., Sokol, K.S. & Izmailova, I.A., *T*-schemes for mathematical simulation of vorticity generation on smooth airfoils in vortex methods. *Herald of the Bauman Moscow State Technical University, Natural Sciences*, **6**(105), 2022. (In Russian.)

[12] Saad, Y., *Iterative Methods for Sparse Linear Systems*, Society for Industrial and Applied Mathematics, 2003.

[13] Karras, T., Maximizing parallelism in the construction of BVHs, octrees, and *k*-d trees. *Proc. Intern. Conf. Eurographics/SIGGRAPH*, pp. 33–37, 2012.

[14] Batista, M. & Karawia, A.A., The use of the Sherman–Morrison–Woodbury formula to solve cyclic block tri-diagonal and cyclic block penta-diagonal linear systems of equations. *Applied Mathematics and Computation*, **210**(2), pp. 558–563, 2009.

[15] Kuzmina, K., Marchevsky, I., Soldatova, I. & Izmailova, Y., On the scope of Lagrangian vortex methods for two-dimensional flow simulations and the POD technique application for data storing and analyzing. *Entropy*, **23**, 118, 2021.

# ON MESHLESS LAGRANGIAN VORTEX METHODS FOR TWO-DIMENSIONAL FLOW SIMULATION AT MODERATE AND HIGH REYNOLDS NUMBERS

YULIA IZMAILOVA & ILIA K. MARCHEVSKY
Bauman Moscow State Technical University, Russia

## ABSTRACT

Vortex particle methods of computational fluid dynamics belong to the class of meshless Lagrangian particle methods, relying on integral representation of velocity field and boundary integral equations method. One of the main advantages of vortex particle methods is connected with the fact that they do not require to reconstruct or deform the mesh, which is important in coupled FSI problems. Vortex methods are widely used to estimate hydrodynamic loads acting on structural elements, especially for essentially unsteady non-potential flow regimes with intensive vortex generation. In a number of engineering applications involving modelling of long structures (bridges, pipelines, buildings), the only way to solve the problem is to use flat cross-section approach. In each cross-section, two-dimensional problem of flow simulation around an airfoil is considered. On one hand, modern models of vortex methods allow for a viscous medium simulating, and on the other hand, they are based on direct approximation of the diffusion term (the Laplacian operator in the Navier–Stokes equation). This makes it possible to correctly simulate the flow around airfoils at low Reynolds numbers. At moderate and high Reynolds numbers, simulation using vortex particle methods remains valid only for airfoils with sharp edges where flow separation takes place. For airfoils with smooth boundaries, the results of simulation do not correspond to experimental data. While the macroscopic flow remains two-dimensional, three-dimensional effects related to turbulence significantly influence the microscale behaviour. Numerical experiments have been performed; the scope of the viscous vortex domain method, implemented by the authors in original code (https://github.com/vortexmethods/VM2D), is estimated. A detailed study of the velocity field in a vortex wake behind a circular cylinder showed that the spatial and temporal spectra of turbulence kinetic energy align well with the Kolmogorov–Obukhov law, but with an exponent of (−3) instead of (−5/3). This observation is consistent with theoretical results for plane flows where three-dimensional effects do not appear at all. The possible ways for developing of two-dimensional models of vortex methods towards including of LES-type turbulence models are discussed.
*Keywords: vortex particle methods, flow simulation around airfoil, high Reynolds numbers, turbulence spectrum, Kolmogorov–Obukhov law.*

## 1 INTRODUCTION

One of the most important problems in aerohydrodynamics is the simulating the interaction between various structural elements and the flow. In engineering applications, the flow of the medium itself is typically not of independent interest. Instead, it is essential to calculate the loads acting on the streamlined airfoil(s), as the magnitude and nature of these loads determine the behaviour of movable or deformable structures. Such problems arise when simulating the flow around structural components, particularly in long-span bridges, various cable structures, overhead power lines, underwater pipelines and hoses, components of aircraft structures, heat exchanger tubes of power plants, etc.

Currently, there are many different approaches to solving computational fluid dynamics problems. The most common methods belong to the mesh-based class, including the finite difference method, the finite volume method, the finite element method and their various modifications. Note that in the flow simulation problems these methods allow for the accurate consideration of many factors relevant to the physical processes being studied. This results

WIT Transactions on Engineering Sciences, Vol 136, 
www.witpress.com, ISSN 1743-3533 (on-line)
doi:10.2495/BE470101

in a broad range of applicability for mesh-based methods, although the computational cost can be rather high, especially when one deals with unsteady flow simulation around moving/deformable airfoils.

Another class, namely Lagrangian meshless methods, includes vortex particle methods [1]–[3], which scope is limited to incompressible flows, but this simplification seems acceptable for many technical applications. In vortex particle methods, vorticity is considered as the primary computational variable. Known vorticity distribution allows for reconstructing the velocity field at any point in the medium using the generalised Biot–Savart law [4], and the pressure field using analogues of the Bernoulli and Cauchy–Lagrange integrals [5]. Vorticity is generated on the airfoil contour line, thereby satisfying the no-slip boundary condition. Note, when simulating external flow using vortex particle methods, there is no need to bound the flow domain, since the boundary condition of perturbations decay at infinity is automatically and exactly satisfied.

Nowadays, vortex particle methods are actively developing and there are modifications that allow for considering three-dimensional and two-dimensional flows. The corresponding algorithms are incommensurate in numerical complexity, but when solving many practical problems, bodies in the flow have close to cylindrical shape with large elongation. Therefore, instead of simulating the entire spatial flow, it is often sufficient to consider one or several two-dimensional flow problems around individual cross-sections (flat cross-sections method [6]). Algorithms designed for problems that require simulation of the flow around flat airfoils remain highly relevant; their main 'competitive advantage' is low computational complexity, and therefore the ability to perform calculations rapidly.

Nevertheless, due to the specific features of vortex particle methods, there are currently only a few their software implementations that are freely available to researchers and engineers (optimally with open source code). One of them is the VM2D code [7], freely available from the github repository at https://github.com/vortexmethods/VM2D. This code is based on the Viscous vortex domains method [2], [8] and some original author's modifications, mainly related to the use of high-precision T-schemes [9] for solving boundary integral equations, as well as some other algorithms. The VM2D code can be used to simulate two-dimensional flows; it allows for solving a wide range of problems, including coupled FSI ones, using the capabilities of modern multiprocessor computers of various architectures (computations on CPU and GPU are available). The aim of this paper is to consider the scope of existing algorithms of vortex particle methods based on the experience of using the VM2D code and to discuss some directions for the future development of this code in particular and vortex methods for modelling 2D flows in general.

## 2 NUMERICAL SIMULATION OF VISCOUS FLOWS IN VM2D CODE

Note that the methods implemented in the VM2D code were applied to the flow simulation around different types of airfoils and for solving coupled FSI problems. Specifically, in Kuzmina et al. [10] the application of vortex particle methods for the simulation of various two-dimensional flows was explored. That paper shows wide range of applicability of the method for different types of airfoils (circular cylinders; airfoils with corner points; wing airfoils).

Some other results obtained using the VM2D code are described in Marchevsky et al. [7], where a comprehensive review of the code's capabilities for simulating two-dimensional incompressible flows using vortex particle methods is presented. That paper highlights the flexibility of VM2D and its applicability to various hydrodynamic tasks (flow simulation around the Savonius rotor, steady-state regime in the channel with backward-facing step, Blasius flow). Also, in Marchevsky et al. [7], Marchevskii et al. [9] and Kuzmina et al. [10]

one can see the accuracy analysis between the simulation results and the reference results, especially in tabular and plot form, together with the error dependence on the level of discretisation.

In the present study let us focus on the issue of accounting for viscosity in two-dimensional flow simulation. Historically, modern versions of vortex particle methods for solving two-dimensional problems trace back to the discrete vortex method [11]–[14] (in some publications, both the term 'discrete vortex method' and 'lumped vortex method' are used). Emerging in the 1950s, this method appears to be one of the earliest methods of computational fluid dynamics. It provides numerical modelling of the flow of an ideal (non-viscous) fluid that is described by the Euler equations; in such a formulation, it is possible to simulate potential or, at least, irrotational flows. Note that it is possible to reproduce much more complex unsteady separated flow regimes around airfoils – this is especially relevant for thin airfoils (plates), wing airfoils and, in general, for airfoils with sharp edges and corner points. The position of the vortex sheet separation is set at the ends of thin plate (one or both, Fig. 1) or at the corner points; the descending vortex sheet that forms the vortex wake, is considered as a thin surface of tangent discontinuity of velocity.

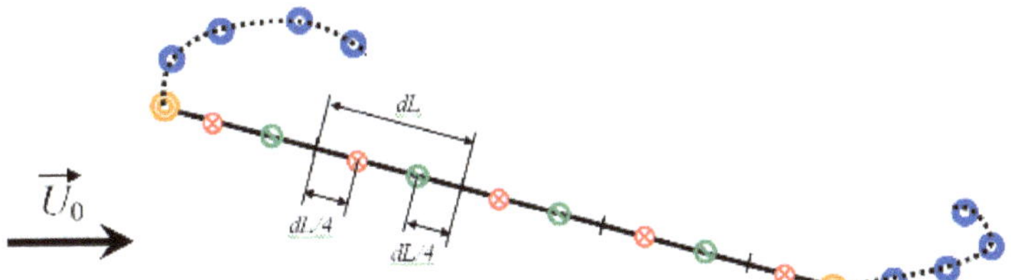


Figure 1: Numerical scheme for flow simulation around a thin plate using the method of discrete vortices with vortex sheets shedding from both ends. Plate splitting into panel is shown as well as positions of control points (red), attached vortices (green), free vortices (orange) and vortex sheets modelled as thin discontinuity surfaces with point vortices.

Although such a model is rather inaccurate, it allows for solving a number of problems of practical interest. Its fundamental issue is the inability to solve problems related to the flow simulation around smooth airfoils, where the position of the flow separation point cannot be specified a priori. It is also impossible to simulate flows at low Reynolds numbers, which are mainly influenced by viscous forces.

The development of vortex particle methods has mainly followed the strategy of taking into account the influence of viscosity when solving the Navier–Stokes equations. Today, the following approaches are known: the random walk method, developed in the 1970s [15], the particle strength exchange (PSE) method, proposed in 1989 [16], the core spreading method, first proposed in 1973 [17] and later developed in 1996 [18], the diffusive velocity method, known in several modifications, and some others. Among these is the method of viscous vortex domains [2], [8], which is implemented in the VM2D code and appears to offer several advantages over the other methods. Numerical experiments using the VM2D code show that flows with a predominance of viscous effects can be simulated quite well: velocity profiles in the Poiseuille flow in a channel, in the Couette flow between two coaxial rotating cylinders (Fig. 2), and in the boundary layer on a thin plate in the Blasius problem (Fig. 3) are reproduced quite accurately, even with a relatively small number of vortex particles.

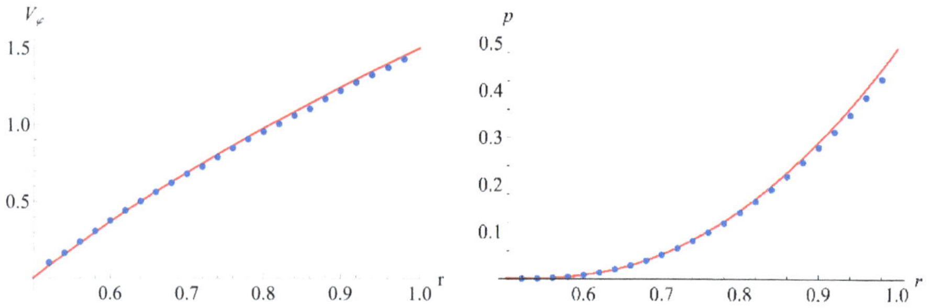


Figure 2: Velocity and pressure profiles for Couette flow between co-rotating cylinders (internal cylinder of radius 0.5 is immovable, external one with radius 1 is rotating); dots indicate numerical solution; solid line represents exact solution against the distance to the axis.

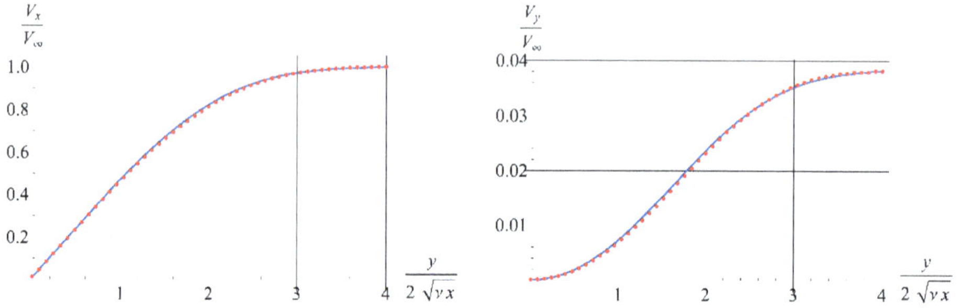


Figure 3: Horizontal $V_x$ and vertical $V_y$ dimensionless velocity profiles in the cross section on a thin plate (Blasius problem); the dependency against the self-similar variable is shown ($\nu$ is kinematic viscosity, $V_\infty$ is incident flow velocity); dots indicate numerical solution; solid line represents exact solution.

Simulating the flow around airfoils at moderate and high Reynolds numbers presents a more complex challenge. On the one hand, the influence of viscous forces is comparatively small (and in the flow domain far from the airfoil boundary is in most cases negligible). However, at the same it remains essential to account for viscosity to accurately model the processes in the near-wall boundary layer, and thus correctly reproduce the separation of the flow from the airfoil contour line.

## 3 UNSTEADY FLOW SIMULATION AROUND CIRCULAR CYLINDER

Given that vortex particle methods are particularly efficient for simulating unsteady and transient flow regimes, let us consider the following model problem of flow simulation around a cylinder that impulsively starts moving with constant speed in still medium. Unsteady drag force acting on the cylinder at different values of the Reynolds number is a 'classical' test problem that has been considered by many researchers. Fig. 4 shows the vortex wakes in the initial phase of the cylinder motion obtained using the VM2D code for two cases that differ only in the values of the viscosity coefficient of the medium. The corresponding values of the Reynolds number are Re = 40 and Re = 200. The ratio of the cylinder diameter to the incident flow velocity is chosen as the time scale.

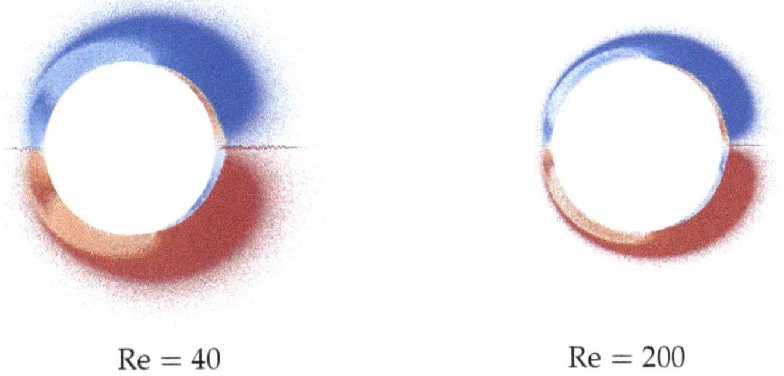


Figure 4: Vortex particles positions in vortex wakes after the impulsively started circular cylinder at Re = 40 and Re = 200 at time step $t_* = 0.5$.

The dependences of the total drag force, caused by the pressure distribution and viscous friction, on the dimensionless time for the considered cases are shown in Fig. 5 in comparison with the results of Bar-Lev and Yang [19], Collins and Dennis [20] and Kousoutsakos and Leonard [21]. It is seen that the simulation results are in good agreement with them.

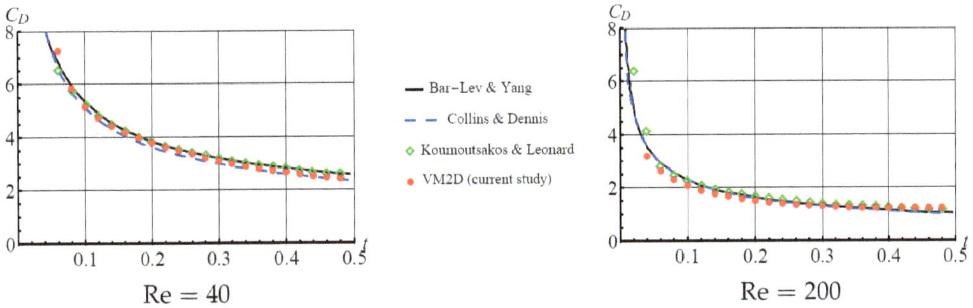


Figure 5: Unsteady drag force coefficient acting on the impulsively started circular cylinder at Re = 40 and Re = 200.

The similar problem was considered for the cylinder velocity corresponding to a higher Reynolds number Re = 3000. The dimensionless simulation time was chosen to be $t_* = 5.0$. The vortex wakes at $t = 1.0$, $t = 2.0$, $t = 3.0$, $t = 4.0$, $t = 5.0$ are shown in Fig. 6.

This problem turns out to be more complicated. To ensure the simulation results aligned with those from other researchers [21]–[27] for Re = 3000, it was necessary to consider a very accurate airfoil discretisation, a small time step, and a sufficiently large number of vortex particles in the flow domain (about half a million). The combination of such high resolution and implementation of the T-schemes for solving the boundary integral equation, made it possible to avoid numerical oscillations of drag and lift forces. These oscillations are common for many known implementations of vortex particle methods. Fig. 7 shows the results for unsteady drag force dependence on time, using the VM2D code in comparison with the other authors results.

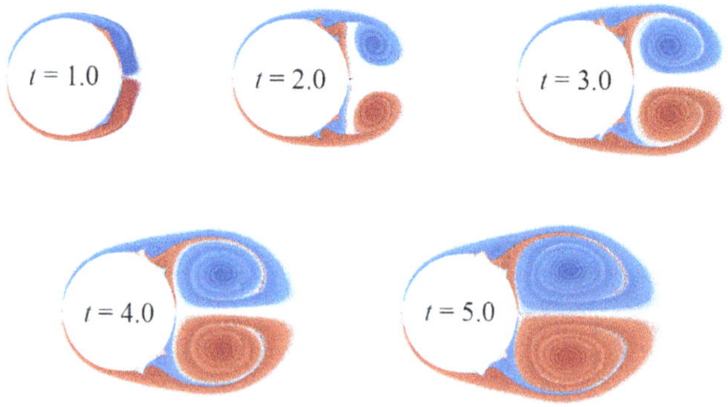


Figure 6: Vortex particles positions in vortex wakes after the impulsively started circular cylinder at Re = 3000 at different time steps.

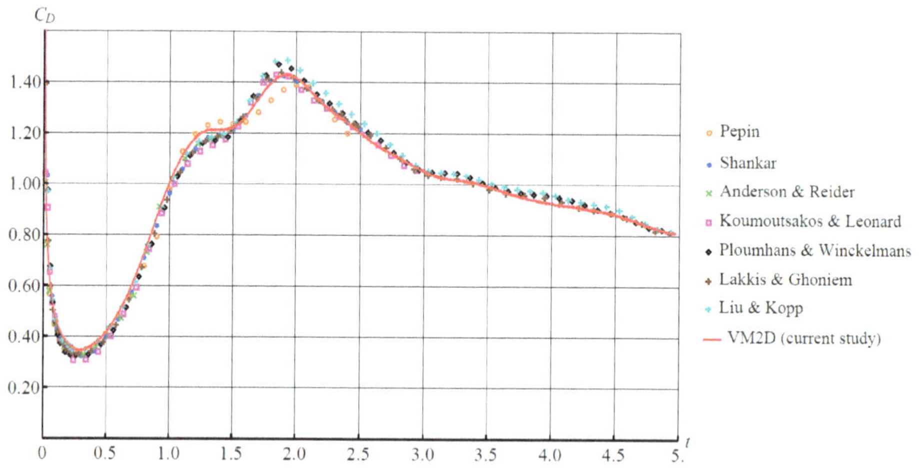


Figure 7: Unsteady drag force coefficient acting the impulsively started circular cylinder at Re = 3000.

Attempts to simulate flows around a circular airfoil at higher Reynolds numbers lead to an undesirable result: when modelling a quasi-steady flow regime using VM2D, the result is qualitatively correct (a Karman vortex street with periodic vortex shedding is observed), but quantitatively incorrect. For example, it is not possible to reproduce the well-known effect of 'stabilisation' of the drag coefficient: it is well-known that averaged drag coefficient of a cylinder remains close to 1.2 in a wide range of Reynolds numbers, and in the region of Re ≈ 1000 even decreases below 1.0 (Fig. 8).

When performing simulation using vortex particle methods, the drag coefficient turns out to be significantly overestimated, close to 1.5…1.6. In addition, it is clear from Fig. 8 that the value $Re = 2 \cdot 10^5$ corresponds to the so-called 'drag crisis', where a sharp decrease in the drag coefficient is observed. Although the qualitative effect of the 'drag crisis' is observed using vortex particle methods (as also noted in Dynnikova [29]), it appears prematurely at approximately $Re = 10^5$.

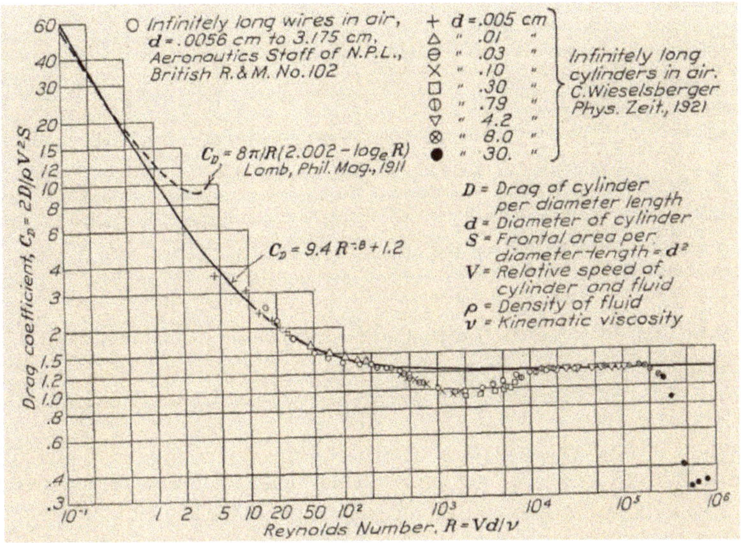


Figure 8: Stationary drag coefficient acting on a circular cylinder in dependence to Reynolds number [28].

## 4 FLOW SIMULATION AROUND WING AIRFOIL

Let us consider the problem of the flow simulation around wing airfoil at moderate Reynolds number values. Fig. 9 shows the vortex wake after a symmetric NACA-0012 airfoil at an angle of incidence of $\alpha = 6°$ in the steady-state flow regime at $\mathrm{Re} = 10^4$ and $\mathrm{Re} = 10^5$. The parameters of simulation were chosen similar to those used for the case of flow around a cylinder considered in the previous section.

Figure 9: Vortex wake after the NACA-0012 airfoil for angle of incidence $\alpha = 6°$ at $\mathrm{Re} = 10^4$ and $\mathrm{Re} = 10^5$.

The dependences of the averaged drag and lift coefficients on the angle of attack (Fig. 10) are in acceptable agreement with the experimental data.

Despite slightly overestimated drag coefficient values, this is not critical due to their smallness. In fact, we can conclude that the flow is simulated correctly even at high Reynolds numbers.

WIT Transactions on Engineering Sciences, Vol 136, © 2024 WIT Press
www.witpress.com, ISSN 1743-3533 (on-line)

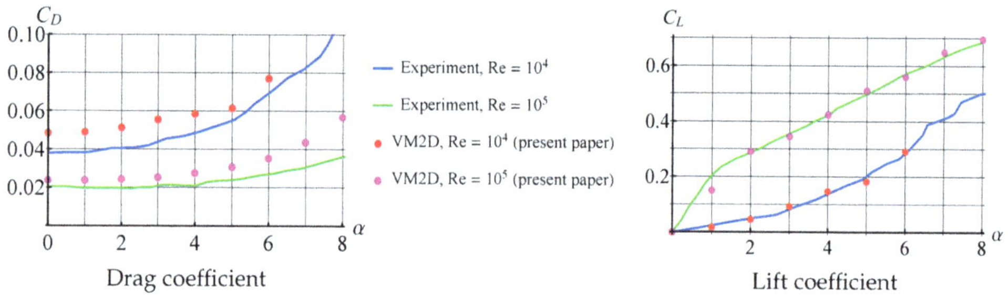


Figure 10: Drag and lift force coefficients against the angle of incidence for NACA-0012 airfoil.

If we continue to increase the angle of incidence of the wing, the computational results in VM2D significantly diverge from the experimental data: mainly, a greatly overestimated value of the lift coefficient is observed. At the same time, as the angle of incidence increases, the flow regime also changes, separation takes place not only at the sharp trailing edge but also on the smooth upper surface of the airfoil. Apparently, these effects are simulated incorrectly.

## 5 ON TURBULENT FLOW SIMULATION USING 2D VORTEX PARTICLE METHODS

Based on the results presented above, the following conclusion can be made. Flow simulation in two-dimensional formulation allows one to obtain correct results in two cases: at low Reynolds numbers – both for smooth airfoils and airfoils with edges and corner points, and at high Reynolds numbers, but only in cases where separation from a smooth surface is not simulated.

Thus, the following hypothesis seems justified: in the mentioned cases, the real (three-dimensional) flow around the structure is in fact essentially flat; at higher Reynolds numbers, the flow can be flat only on the macroscales, while microscale flows turn out to be essentially three-dimensional, but this microscale three-dimensionality affects the macroscopic flow characteristics. This effect is well known, and widespread mesh-based methods reproduce it using the so-called turbulence models, within the framework of the RANS or LES approaches. The averaged (filtered) flow is flat, while the pulsating component of the velocity field is three-dimensional, but not explicitly resolved.

It is well-known that the introduction of special closure models, which effectively embody the specifics of a given turbulence model, arises from the inability to provide the necessary spatial resolution of the computational mesh. If one performs direct numerical simulation, the typical spatial scale of structures that must be resolved in the flow domain is called the Kolmogorov scale and is inversely proportional to the value of $Re^{3/4}$. In discussing small-scale effects, it should be noted that in relatively simple models, the turbulence is assumed to be isotropic in three spatial directions. 'Turbulence' itself is defined as essentially three-dimensional unsteady motion. Due to the stretching of vortices, a distribution of velocity pulsations occurs in the range of wavelengths, from the minimum, determined by the Kolmogorov scale, to the maximum determined by the size of the flow domain. The spectral distribution of the kinetic energy of turbulence has a so-called inertial range (intermediate between the range of large energy-containing vortices and the dissipation range, where the energy of vortices is transferred into heat as a result of the action of viscous forces). In inertial range turbulence can be considered homogeneous and isotropic, and according to the Kolmogorov–Obukhov law, the spectral density of the kinetic energy of turbulence $E$, is proportional to $k^{5/3}$, where $k$ is the wave number [30].

In vortex particle methods for 2D flows, 'direct' reproduction of the mentioned effects is impossible because the primary computational variable, i.e., vorticity is a priori taken to be orthogonal to the flow domain. This, in fact, excludes the possibility of modelling the process of vortex stretching. Nevertheless, if one performs derivation of the estimation for energy spectrum (keeping in mind two-dimensionality of the flow), the inertial range consists of two subregions in 'two-dimensional turbulence': for large wave numbers, the asymptotics $E \sim k^{-3}$ is valid, and the law $E \sim k^{-5/3}$ applies only in a relatively narrow range [31]. This dependence can be practically tested for flow simulation using the VM2D code.

As an example, a simulation of the flow after a circular airfoil was performed at a Reynolds number $\mathrm{Re} = 10^6$. The value of the steady-state (averaged) drag coefficient is close to 0.45 (see the issue of the drag crisis discussed above). The plots of the spectral density of the kinetic energy of turbulence are shown in Fig. 11. The left plot was obtained by direct processing the simulation results: the velocity of the medium was calculated on a uniform grid with a small step in square area. The side length of the square was equal to half the radius of the circle and it was located behind it at a distance equal to 1/20 of the radius of the cylinder. Then two-dimensional discrete Fourier transform was performed; the plot shows the squared absolute values of the Fourier coefficients against the wave vector modulus; the plot is shown in a double logarithmic scale. The right plot was obtained by processing the time dependence for the value of velocity at a fixed point placed at the distance of 1/15 of the radius at angle 120°. A discrete Fourier transform was performed, the value $k = f/U$ was plotted along the horizontal axis, where $f$ is the frequency, $U$ is the incident flow velocity, and the squared absolute values of the corresponding Fourier coefficient were plotted along the vertical axis.

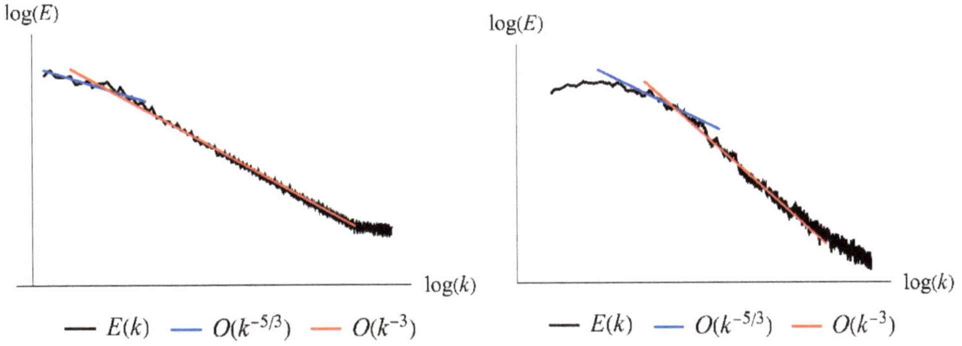


Figure 11: Spectral density plots of turbulence kinetic energy. (Left) For spatial velocity field at fixed time; and (Right) For velocity dependency against time at fixed point).

It is seen that the kinetic energy density spectrum of 'two-dimensional turbulence' is correctly reproduced in the inertial range without involving any additional models. This allows us to conclude that the algorithms for 2D flow simulation implemented in VM2D are sufficiently effective. Attempts to achieve a similar result using mesh-based methods (the finite volume method implemented in OpenFOAM) without involving turbulence models were unsuccessful. Such attempts would require an unacceptably tiny grid and time step, leading to prohibitively high computational costs.

Currently, only a few studies have attempted to combine vortex methods with RANS and LES approaches for simulating turbulent flows [32]–[35]. However, such approaches are not widespread, at least for near-wall flows simulation. This can be partly attributed to the challenges in accurately solving the boundary integral equation for the vortex sheet intensity generated on the airfoil, a problem that persisted until recently. Actually, this significantly limited the researchers in achieving high resolution in the near-wall region. The use of the above-mentioned *T*-schemes allows us to solve this problem and opens up the possibility of developing new modifications of vortex particle methods, including the possibility of turbulent effects simulation.

However, we would like to clarify that the primary goal of this work is to observe the problem and highlight the significant influence these 3D effects can have on flow simulations. This study aims to provide an initial exploration and set the stage for more detailed analyses. Quantitative comparisons and a thorough assessment of the extent of this influence fall beyond the scope of the current paper and warrant a dedicated, comprehensive investigation. Such a study is essential and we plan to address it in a separate follow-up work, which will focus specifically on the quantification and detailed analysis of these effects.

## 6 CONCLUSION

Using the VM2D code, the possibility of applying vortex methods based on the viscous vortex domain method to solving model problems on simulating flows around airfoil at moderate and high Reynolds numbers is considered. At low Reynolds numbers, the simulation results are in good agreement with experimental data. However, significant errors are observed at high Reynolds numbers.

For small angles of incidence, the presented dependencies of drag and lift forces on the airfoil at Reynolds numbers $\text{Re} = 10^4$ and $\text{Re} = 10^5$ are close to the experimental results. However, as the angle of incidence increases, discrepancies emerge again. This can be explained by the fact that at higher angles, flow separation takes place not only on the sharp edge, but also on the upper smooth surface of the airfoil, and such separation as in the case of a circular airfoil, is simulated incorrectly. The latter is due to the fact that at high Reynolds numbers the flow can be modelled as flat, but only on macroscales, while microscale flow is essentially three-dimensional, especially in the zone of separation. Such microscale three-dimensionality cannot be reproduced in two-dimensional algorithms of vortex particle methods, but it has a significant impact on the flow.

At the same time, analysis of the spectrum of the kinetic energy of turbulence shows that it corresponds to 'two-dimensional turbulence' with high accuracy. Thus, we can conclude that the algorithms for modelling flat flows implemented in VM2D are sufficiently effective.

## REFERENCES

[1] Cottet, G.-H. & Koumoutsakos, P.D., *Vortex Methods: Theory and Practice*, Cambridge University Press, 328 pp., 2000.

[2] Andronov, P.R., Guvernyuk, S.V. & Dynnikova, G.Y., *Vortex Methods for Non-Stationary Hydrodynamic Loads Estimation*, Moscow State University, 184 pp., 2006.

[3] Mimeau, C. & Mortazavi, I., A review of vortex methods and their applications: from creation to recent advances. *Fluids*, **6**(2), pp. 1–25, 2021.

[4] Kempka, S.N., Glass, M.W., Peery, J.S., Strickland, J.H. & Ingber, M.S., Accuracy considerations for implementing velocity boundary conditions in vorticity formulations. *SANDIA Report*, SAND96-0583, UC-700, 52 pp., 1996.

[5] Dynnikova, G.Y., An analog of the Bernoulli and Cauchy–Lagrange integrals for a time dependent vortex flow of an ideal incompressible fluid. *Fluid Dynamic*, **35**(1), pp. 31–41, 2000.

[6] Devnin, S.I., *Hydroelasticity of Structures under Separated Flow*, Sudostroenie: Leningrad, 99 pp., 1975.

[7] Marchevsky, I., Sokol, K., Ryatina, E. & Izmailova, Y., The VM2D open source code for two-dimensional incompressible flow simulation by using fully Lagrangian vortex particle methods. *Axioms*, **12**(3), pp. 1–24, 2023.

[8] Dynnikova, G.Y., The Lagrangian approach to solving the time-dependent Navier–Stokes equations. *Doklady Physics*, **49**, pp. 648–652, 2004.

[9] Marchevskii, I.K., Sokol, K.S. & Izmailova, Y.A. T-schemes for mathematical modelling of vorticity generation on smooths airfoils in vortex particle methods. *Herald of the Bauman Moscow State Technical University, Series Natural Sciences*, **6**(105), pp. 33–59, 2022.

[10] Kuzmina, K., Marchevsky, I., Soldatova, I. & Izmailova, Y., On the scope of Lagrangian vortex methods for two-dimensional flow simulations and the POD technique application for data storing and analysing. *Entropy*, **23**(1), pp. 1–17, 2021.

[11] Belotserkovsky, S.M. & Lifanov, I.K., *Method of Discrete Vortices*, CRC Press: Boca Raton, 464 pp., 1992.

[12] Lifanov, I.K., *Singular Integral Equations and Discrete Vortices*, VSP: Utrecht, The Netherlands, 475 pp., 1996.

[13] Katz, J. & Plotkin, A., *Low-Speed Aerodynamics: From Wing Theory to Panel Methods*, McGraw-Hill: Singapore, 632 pp., 1991.

[14] McBain, G.D., *Theory of Lift: Introductory Computational Aerodynamics in MATLAB/Octave*, Wiley: New York, 342 pp., 2012.

[15] Chorin, A.J., Numerical study of slightly viscous flow. *Journal of Fluid Mechanics*, **57**(4), pp. 785–796, 1973.

[16] Degond, P. & Mas-Gallic, S., The weighted particle method for convection-diffusion equations. I. The case of an isotropic viscosity. *Mathematics of Computation*, **53**(188), pp. 485–507, 1989.

[17] Kuwahara, K. & Takami, H., Numerical studies of two-dimensional vortex motion by a system of point vortices. *Journal of the Physical Society of Japan*, **34**(1), pp. 247–253, 1973.

[18] Rossi, L.F., Resurrecting core spreading vortex methods: A new scheme that is both deterministic and convergent. *SIAM Journal on Scientific Computing*, **17**(2), pp. 370–397, 1996.

[19] Bar-Lev, M. & Yang, H.T., Initial flow over an impulsively started circular cylinder. *Journal of Fluid Mechanics*, **72**(4), pp. 625–647, 1975.

[20] Collins, W.M. & Dennis, S.C.R., The initial flow past an impulsively started circular cylinder. *Quarterly Journal of Mechanics and Applied Mathematics*, **26**(1), pp. 53–75, 1973.

[21] Koumoutsakos, P. & Leonard, A., High-resolution simulations of the flow around an impulsively started cylinder using vortex methods. *Journal of Fluid Mechanics*, **296**, pp. 1–38, 1995.

[22] Pepin, F.M., Simulation of the flow past an impulsively started cylinder using a discrete vortex method. PhD thesis, California Institute of Technology, Pasadena, USA, 1990.
[23] Shankar, S., A new mesh-free vortex method. PhD thesis, FAMU-FSU College of Engineering, Tallahassee, USA, 1996.
[24] Anderson, C.B. & Reider, M.B., A high order explicit method for the computation of flow about a circular cylinder. *Journal of Computational Physics*, **125**(1), pp. 207–224, 1996.
[25] Ploumhans, P. & Winckelmans, G.S., Vortex methods for high-resolution simulations of viscous flow past bluff bodies of general geometry. *Journal of Computational Physics*, **165**(2), pp. 354–406, 2000.
[26] Lakkis, I. & Ghoniem, A., A high resolution spatially adaptive vortex method for separating flows. Part I: Two-dimensional domains. *Journal of Computational Physics*, **228**, pp. 491–515, 2009.
[27] Liu, Z. & Kopp, G.A., High-resolution vortex particle simulations of flows around rectangular cylinders. *Computers and Fluids*, **40**(1), pp. 2–11, 2011.
[28] Zahm, A.F., Flow and drag formulas for simple quadrics. NACA Technical Report No. 253, 23 pp., 1927.
[29] Dynnikova, G.Y., Fast technique for solving the N-body problem in flow simulation by vortex methods. *Computational Mathematics and Mathematical Physics*, **49**(8), pp. 1389–1396, 2009.
[30] Kolmogorov, A.N., The local structure of turbulence in incompressible viscous fluid for very large Reynolds numbers. *Proceedings of the Royal Society A: Mathematical, Physical and Engineering Sciences*, pp. 9–13, 1991.
[31] Kraichnan, R.H., Inertial ranges in two-dimensional turbulence. *Physics of Fluids*, **10**(7), pp. 1417–1423, 1967.
[32] Pereira, L.A.A., Hirata, H. & Silveira Neto, A., Vortex method with turbulence sub-grid scale modelling. *Journal of the Brazilian Society of Mechanical Sciences and Engineering*, **25**(2), pp. 140–146, 2003.
[33] Yokota, R. & Shinnosuke, O., Vortex methods for the simulation of turbulent flows: Review. *Journal of Fluid Science and Technology*, **6**(1), pp. 14–29, 2011.
[34] Branlard, E., Papadakis, G., Gaunaa, M., Winckelmans, G. & Larsen T., Aeroelastic large eddy simulations using vortex methods: unfrozen turbulent and sheared inflow. *Journal of Physics: Conference Series*, **625**, 012019, 2015.
[35] Alvarez, E.J. & Ning, A., Reviving the vortex particle method: A stable formulation for meshless large eddy simulation. arXiv preprint arXiv:2206.03658, 2022.

# SECTION 4
# MESHLESS AND MULTISCALE PROBLEMS

# ACOUSTIC WAVE SCATTERING BY NULL-THICKNESS BODIES WITH COMPLEX GEOMETRY

ANTONIO ROMERO[1], ROCIO VELÁZQUEZ-MATA[1], ANTONIO TADEU[2,3] & PEDRO GALVÍN[1,4]
[1]Escuela Técnica Superior de Ingeniería, Universidad de Sevilla, Spain
[2]Department of Civil Engineering, University of Coimbra, CERIS, Portugal
[3]Itecons, Institute of Research and Technological Development in Construction, Energy, Environment and Sustainability, Portugal
[4]ENGREEN, Laboratory of Engineering for Energy and Environmental Sustainability, Universidad de Sevilla, Spain

ABSTRACT

This paper proposes a general formulation of the BEM based on the Burton–Miller method to study scattering wave propagation by null-thickness bodies with complex geometry. This approach allows the use of arbitrary high-order elements and exact boundary geometry. We use the Bézier–Bernstein form of a polynomial as an approximation basis to represent both geometry and field variables. The solution of the element interpolation problem in the Bézier–Bernstein space defines generalised Lagrange interpolation functions that are used as element shape functions. The proposed procedure consists of a new quadrature rule for the accurate evaluation of integral kernels in the sense of the Cauchy principal and the Hadamard finite part by an exclusively numerical procedure.

*Keywords: hypersingular formulation, dual BEM, boundary integral equation, hypersingular kernels, singular kernels.*

## 1 INTRODUCTION

Krishnasamy et al. [1], Liu and Rizzo [2] and Liu [3] developed dual formulations using the Burton–Miller method to study wave scattering from thin shapes. The Burton–Miller method has been widely used to avoid spurious eigenfrequencies in exterior acoustic problems. The method is based on a linear combination of the boundary integral equation (BIE) and the hipersingular boundary integral equation (HBIE) which is typically written as BIE $+ \mu$HBIE $= 0$, where $\mu$ is the coupling parameter. Although this approach is much simpler and more general than other dual formulations, it has not been widely applied in wave propagation from null-thickness bodies.

## 2 DUAL BEM FORMULATIONS

This work analyses the wave propagation in a fluid medium in presence of thin rigid bodies with arbitrary shape (Fig. 1) according to the two-dimensional Helmholtz equation:

$$\nabla^2 u(\mathbf{x}) + \kappa^2 u(\mathbf{x}) = 0, \qquad \mathbf{x} \in \Omega, \tag{1}$$

where $u(\mathbf{x})$ is the velocity potential and $\kappa > 0$ denotes the wavenumber. The system is elicited by an incident wavefield caused by a point source defined by $u_0(\mathbf{x}, \mathbf{x}_0) = -\iota H_0^{(1)}(\kappa r_0)/2$, where $H_0^{(1)}$ is the Hankel function of first kind, $r_0 = ||\mathbf{x} - \mathbf{x}_0||$ is the distance to the source point $\mathbf{x}_0$, and the unit imaginary number is denoted by the Greek letter $\iota$ to prevent confusion with some subscripts used in the paper.

The problem definition is completed by setting the boundary conditions for on $\Gamma = \partial\Omega$. Dirichlet and Neumann boundary conditions are defined as $u(\mathbf{x}) = \overline{u}(\mathbf{x})$ and $q(\mathbf{x}) = \partial u(\mathbf{x})/\partial \mathbf{n}(\mathbf{x}) = \overline{q}(\mathbf{x})$, respectively, where $\mathbf{n}(\mathbf{x})$ is the outward normal at the boundary. Also, it is possible to adopt the Robin boundary condition to gather the three types of boundary

WIT Transactions on Engineering Sciences, Vol 136, © 2024 WIT Press
www.witpress.com, ISSN 1743-3533 (on-line)
doi:10.2495/BE470111

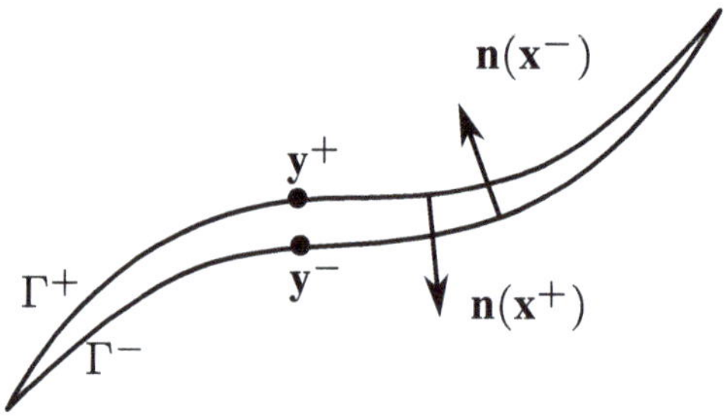


Figure 1: Boundary definition of a thin body.

conditions in one expression:

$$\alpha(\mathbf{x})u(\mathbf{x}) + \beta(\mathbf{x})q(\mathbf{x}) = \gamma(\mathbf{x}), \qquad \mathbf{x} \in \Gamma, \tag{2}$$

where $\alpha(\mathbf{x})$, $\beta(\mathbf{x})$ and $\gamma(\mathbf{x})$ are known parameters [4].

The Green's function $\Psi(\mathbf{x}, \mathbf{y})$ for Helmholtz equation in an unbounded region at receiver position $\mathbf{x}$ due to a source acting at $\mathbf{y}$ is the solution to:

$$\nabla^2\Psi(\mathbf{x}, \mathbf{y}) + \kappa^2\Psi(\mathbf{x}, \mathbf{y}) = -\delta(r) \tag{3}$$

where $\delta$ is the Dirac delta function and $r = ||\mathbf{x} - \mathbf{y}||$. The solution to this equation defines the Green's function for an unbounded region in the frequency domain:

$$\Psi(\mathbf{x}, \mathbf{y}) = -\frac{\iota}{4} H_0^{(1)}(\kappa r) \tag{4}$$

It is assumed that $\kappa^2$ is not an eigenvalue of $\nabla^2 u(\mathbf{x}) + \kappa^2 u(\mathbf{x}) = 0$ subject to the homogeneous form of the imposed boundary conditions.

## 2.1 Indeterminacy of the boundary integral equation

This section briefly reviews the degeneracy of the boundary integral equation (BIE) as the thickness of a thin body tends to zero. The BIE for the velocity potential can be written as follows [5]:

$$c(\mathbf{y})u(\mathbf{y}) = u_0(\mathbf{y}) + \int_\Gamma \left( \frac{\partial u(\mathbf{x})}{\partial \mathbf{n}(\mathbf{x})} \Psi(\mathbf{x}, \mathbf{y}) - u(\mathbf{x}) \frac{\partial \Psi(\mathbf{x}, \mathbf{y})}{\partial \mathbf{n}(\mathbf{x})} \right) d\Gamma \tag{5}$$

where $\mathbf{y}$ is the collocation point and the integral-free term $c(\mathbf{y})$ depends only on the boundary geometry. The discretised form of eqn (5) when the boundary is discretised into $N$ elements with $\Gamma = \bigcup_{j=1}^{N} \Gamma^j$ is:

$$c(\mathbf{y})u(\mathbf{y}) = u_0(\mathbf{y}) + \sum_{j=1}^{N} \int_{\Gamma^j} \left( \frac{\partial u(\mathbf{x})}{\partial \mathbf{n}(\mathbf{x})} \Psi(\mathbf{x}, \mathbf{y}) - u(\mathbf{x}) \frac{\partial \Psi(\mathbf{x}, \mathbf{y})}{\partial \mathbf{n}(\mathbf{x})} \right) d\Gamma \tag{6}$$

Then, the field variables within an element $\Gamma^j$ are interpolated from the nodal values $u^i$ using element shape functions $\phi^i(\mathbf{x})$ of order $p$:

$$u(\mathbf{x}) = \sum_{i=0}^{p} \phi^i(\mathbf{x}) u^i \tag{7}$$

The element approximation is substituted into eqn (6) to yield the following expression:

$$c(\mathbf{y})u(\mathbf{y}) = u_0(\mathbf{y}) + \sum_{j=1}^{N} \left\{ \sum_{i=0}^{p} \left[ \left( \int_{\Gamma^j} \phi^i(\mathbf{x}) \Psi(\mathbf{x}, \mathbf{y}) \, d\Gamma \right) q^i - \left( \int_{\Gamma^j} \phi^i(\mathbf{x}) \frac{\partial \Psi(\mathbf{x}, \mathbf{y})}{\partial \mathbf{n}(\mathbf{x})} \, d\Gamma \right) u^i \right] \right\} \tag{8}$$

which is the classical BEM formulation. This expression for a collocation point $\mathbf{y}^+$ (see Fig. 1) can be further elaborated by separating the element integral according to $\Gamma^+$ and $\Gamma^-$ (a similar equation is obtained for collocation in $\mathbf{y}^-$):

$$\begin{aligned} c(\mathbf{y}^+)u(\mathbf{y}^+) &= u_0(\mathbf{y}) \\ &+ \sum_{\Gamma^j \in \Gamma^+} \left\{ \sum_{i=0}^{p} \left[ \left( \int_{\Gamma^j} \phi^i(\mathbf{x}) \Psi(\mathbf{x}, \mathbf{y}^+) \, d\Gamma \right) q^i - \left( \int_{\Gamma^j} \phi^i(\mathbf{x}) \frac{\partial \Psi(\mathbf{x}, \mathbf{y}^+)}{\partial \mathbf{n}(\mathbf{x}^+)} \, d\Gamma \right) u^i \right] \right\} \\ &+ \sum_{\Gamma^j \in \Gamma^-} \left\{ \sum_{i=0}^{p} \left[ \left( \int_{\Gamma^j} \phi^i(\mathbf{x}) \Psi(\mathbf{x}, \mathbf{y}^+) \, d\Gamma \right) q^i - \left( \int_{\Gamma^j} \phi^i(\mathbf{x}) \frac{\partial \Psi(\mathbf{x}, \mathbf{y}^+)}{\partial \mathbf{n}(\mathbf{x}^-)} \, d\Gamma \right) u^i \right] \right\} \end{aligned} \tag{9}$$

Thus, the element nodes are chosen as collocation points to yield a non-symmetric linear system of equations relating the nodal variables. The system of equations is rearranged according to the boundary conditions.

For example, the system of equations for a fixed scatter with null-thickness and sound hard condition, $q(\mathbf{x}) = 0$, $\mathbf{x} \in \Gamma$, can be written as follows:

$$\begin{bmatrix} \frac{1}{2}\mathbf{I} + \mathbf{H}^{++} & -\mathbf{H}^{++} \\ \mathbf{H}^{++} & \frac{1}{2}\mathbf{I} - \mathbf{H}^{++} \end{bmatrix} \begin{Bmatrix} \mathbf{u}^+ \\ \mathbf{u}^- \end{Bmatrix} = \begin{Bmatrix} \mathbf{u}_0^+ \\ \mathbf{u}_0^- \end{Bmatrix} \tag{10}$$

where, $\mathbf{u}^+$, $\mathbf{u}^-$, $\mathbf{u}_0^+$ and $\mathbf{u}_0^-$ correspond to nodal values of $u^i$ and $u_0$ at the boundaries $\Gamma^+$ and $\Gamma^-$. The boundary is assumed to be smooth, i.e. $c(\mathbf{y}^+) = 1/2$. $\mathbf{H}^{++}$ is the influence matrix obtained from the element integration in $\Gamma^+$ (second and fourth integrals in eqn (9)) considering that the Green's function satisfies:

$$\frac{\partial \Psi(\mathbf{x}, \mathbf{y}^+)}{\partial \mathbf{n}(\mathbf{x}^+)} = -\frac{\partial \Psi(\mathbf{x}, \mathbf{y}^+)}{\partial \mathbf{n}(\mathbf{x}^-)} \tag{11}$$

Obviously, the determinant of the coefficient matrix in eqn (10) is equal to zero and the system degenerates.

Similar occurs with the hypersingular boundary integral equation (HBIE). Taking the normal derivative in eqn (9) at the collocation point and assuming a smooth boundary gives:

$$\begin{aligned} \frac{1}{2}\frac{\partial u(\mathbf{y}^+)}{\partial \mathbf{n}(\mathbf{y}^+)} &= \frac{\partial u_0(\mathbf{y}^+)}{\partial \mathbf{n}(\mathbf{y}^+)} \\ &+ \sum_{\Gamma^j \in \Gamma^+} \left\{ \sum_{i=0}^{p} \left[ \left( \int_{\Gamma^j} \phi^i(\mathbf{x}) \frac{\partial \Psi(\mathbf{x}, \mathbf{y}^+)}{\partial \mathbf{n}(\mathbf{y}^+)} \, d\Gamma \right) q^i - \left( \int_{\Gamma^j} \phi^i(\mathbf{x}) \frac{\partial^2 \Psi(\mathbf{x}, \mathbf{y}^+)}{\partial \mathbf{n}(\mathbf{x}^+) \partial \mathbf{n}(\mathbf{y}^+)} \, d\Gamma \right) u^i \right] \right\} \\ &+ \sum_{\Gamma^j \in \Gamma^-} \left\{ \sum_{i=0}^{p} \left[ \left( \int_{\Gamma^j} \phi^i(\mathbf{x}) \frac{\partial \Psi(\mathbf{x}, \mathbf{y}^+)}{\partial \mathbf{n}(\mathbf{y}^+)} \, d\Gamma \right) q^i - \left( \int_{\Gamma^j} \phi^i(\mathbf{x}) \frac{\partial^2 \Psi(\mathbf{x}, \mathbf{y}^+)}{\partial \mathbf{n}(\mathbf{x}^-) \partial \mathbf{n}(\mathbf{y}^+)} \, d\Gamma \right) u^i \right] \right\} \end{aligned} \tag{12}$$

where

$$\frac{\partial^2\Psi(\mathbf{x}^+,\mathbf{y}^+)}{\partial\mathbf{n}(\mathbf{x}^+)\partial\mathbf{n}(\mathbf{y}^+)} = -\frac{\partial\Psi(\mathbf{x}^-,\mathbf{y}^+)}{\partial\mathbf{n}(\mathbf{x}^-)\partial\mathbf{n}(\mathbf{y}^+)} \tag{13}$$

Then, the system of equation for a fixed scatter with null-thickness and sound hard condition degenerates as in the BIE:

$$\begin{bmatrix} \widetilde{\mathbf{H}}^{++} & -\widetilde{\mathbf{H}}^{++} \\ -\widetilde{\mathbf{H}}^{++} & \widetilde{\mathbf{H}}^{++} \end{bmatrix} \begin{Bmatrix} \mathbf{u}^+ \\ \mathbf{u}^- \end{Bmatrix} = \begin{Bmatrix} \mathbf{q}_0^+ \\ \mathbf{q}_0^- \end{Bmatrix} \tag{14}$$

where $\mathbf{q}_0^+$ and $\mathbf{q}_0^-$ represent the normal derivatives of the point source, $\partial u_0(\mathbf{y}^+)/\partial\mathbf{n}(\mathbf{y}^+)$ and $\partial u_0(\mathbf{y}^-)/\partial\mathbf{n}(\mathbf{y}^-)$, respectively, and the influence matrix $\widetilde{\mathbf{H}}^{++}$ is obtained from element integration.

### 2.2 Dual formulations

In practice, dual formulations based on the combination of the BIE and HBIE are used to define a well-posed system of equations [1].

The most common procedure is to collocate in a single surface of the scatter ($\Gamma^+$ or $\Gamma^-$) using both the BIE and the HBIE according to eqns (9) and (12):

$$\begin{aligned} \frac{1}{2}\Sigma u(\mathbf{y}^+) = u_0(\mathbf{y}^+) + \sum_{\Gamma^j\in\Gamma^+} \Bigg\{ \sum_{i=0}^{p} \Bigg[ \left( \int_{\Gamma^j} \phi^i(\mathbf{x})\Psi(\mathbf{x},\mathbf{y}^+)\, d\Gamma \right) \Sigma q^i \\ - \left( \int_{\Gamma^j} \phi^i(\mathbf{x}) \frac{\partial\Psi(\mathbf{x},\mathbf{y}^+)}{\partial\mathbf{n}(\mathbf{x}^+)}\, d\Gamma \right) \Delta u^i \Bigg] \Bigg\} \end{aligned} \tag{15}$$

and

$$\begin{aligned} \frac{1}{2}\Delta q(\mathbf{y}^+) = \frac{\partial u_0(\mathbf{y}^+)}{\partial\mathbf{n}(\mathbf{y}^+)} + \sum_{\Gamma^j\in\Gamma^+} \Bigg\{ \sum_{i=0}^{p} \Bigg[ \left( \int_{\Gamma^j} \phi^i(\mathbf{x}) \frac{\partial\Psi(\mathbf{x},\mathbf{y}^+)}{\partial\mathbf{n}(\mathbf{y}^+)}\, d\Gamma \right) \Sigma q^i \\ - \left( \int_{\Gamma^j} \phi^i(\mathbf{x}) \frac{\partial^2\Psi(\mathbf{x},\mathbf{y}^+)}{\partial\mathbf{n}(\mathbf{x}^+)\partial\mathbf{n}(\mathbf{y}^+)}\, d\Gamma \right) \Delta u^i \Bigg] \Bigg\} \end{aligned} \tag{16}$$

where $\Sigma u(\mathbf{y}^+) = u(\mathbf{y}^+) + u(\mathbf{y}^-)$ and $\Delta q(\mathbf{y}^+) = q(\mathbf{y}^+) - q(\mathbf{y}^-)$. Similarly, the increment of potential $\Delta u^i$ and the total flux and $\Sigma q^i$ are defined at the nodal point $i$. The system of equations for a fixed scatter can be written as:

$$\begin{bmatrix} \frac{1}{2}\mathbf{I} & \mathbf{H}^{++} \\ \mathbf{0} & \widetilde{\mathbf{H}}^{++} \end{bmatrix} \begin{Bmatrix} \Sigma\mathbf{u} \\ \Delta\mathbf{u} \end{Bmatrix} = \begin{Bmatrix} \mathbf{u}_0^+ \\ \mathbf{q}_0^+ \end{Bmatrix} \tag{17}$$

This procedure is commonly referred to as the *dual formulation*. The HBIE equations in (17) can also be used alone if only the increment of the potential $\Delta\mathbf{u}$ needs to be calculated. It should be noted that this approach is only valid for null-thickness scatters and should be further elaborated in presence of thick bodies [6].

Krishnasamy et al. [1], Liu and Rizzo [2] and Liu [3] developed dual formulations using the Burton–Miller method to study wave scattering from thin shapes. The Burton–Miller method has been widely used to avoid spurious eigenfrequencies in exterior acoustic problems. The method is based on a linear combination of the boundary integral equation (BIE) and the hipersingular boundary integral equation (HBIE) which is typically written as BIE + $\mu$HBIE $= 0$, where $\mu$ is the coupling parameter. Although this approach is much simpler and more general than other dual formulations, it has not been widely applied in wave propagation from null-thickness bodies.

A general formulation of the BEM based on the Burton–Miller method is then proposed to represent acoustic wave propagation problems involving complex thin and thick geometries. The approach presented herein allows the use of arbitrary high-order elements and exact boundary geometry. The proposed method improves the versatility of the BEM as both the BIE and HBIE are evaluated numerically.

### 2.3 Treatment of singular and hypersingular kernels

The element integrals in eqns (9) and (12) are typically of the form:

$$I = \int_{\Gamma^j} \phi^i(\mathbf{x})\mathcal{F}(\mathbf{x}, \mathbf{y})\, d\Gamma(\mathbf{x}) \tag{18}$$

There is an inherent difficulty in boundary element formulations related to the treatment of integral kernels, in particular when the collocation node belongs to the integration element. These integrals are classified as weakly-singular, singular and hypersingular and should be understood in the sense of the Cauchy Principal Value (CPV) or in the Hadamard Finite Part (FP). Combined formulations such as the above become more difficult as all these types of integrals appear, which are typically treated with specific techniques that also depend on the element order. The lack of generality is therefore one of the main drawback on dual formulations.

Kolm and Rokhlin [7] pointed out the need to develop integration schemes that allow evaluation without looking at the integral kernels. The authors presented an algorithm to construct quadrature rules for the simultaneous evaluation of the aforementioned integrals in order to obtain simplified implementations. In this regard, we proposed in a previous work [8] a procedure to design quadrature rules for weakly-singular and singular integrals. Following, the quadrature rules are extended to hypersingular integrals.

The quadrature rules in Velázquez-Mata et al. [8] are derived from the fact that the element shape function $\phi^i(\mathbf{x})$ can be represented in the Bernstein form as [9]:

$$\phi^i(t) = \sum_{k=0}^{n} c_k^i B_k^n(t)\,, \quad i = 0, \ldots, p \tag{19}$$

where $c_k^i$ are control points and $B_k^n(t)$ is the Bernstein polynomial of order $n$ defined over the interval $t \in [0, 1]$:

$$B_k^n(t) = \binom{n}{k} t^k (1-t)^{n-k}, \quad k = 0, \ldots, n \tag{20}$$

Then, the integral kernel in eqn (18) can be rewritten as follows in the natural coordinate $\xi$:

$$\begin{aligned} I = \int_{-1}^{1} \phi^i(\xi)\mathcal{F}(\xi,\mathbf{y})\frac{d\Gamma}{d\xi}\,d\xi &= \int_0^1 \phi^i(t)\mathcal{F}(t,\mathbf{y})\frac{d\Gamma}{d\xi}\frac{d\xi}{dt}\,dt \\ &= \int_0^1 \left(\sum_{k=0}^{n} c_k^i B_k^n(t)\right)\mathcal{F}(t,\mathbf{y})\frac{d\Gamma}{d\xi}\frac{d\xi}{dt}\,dt \\ &= \sum_{k=0}^{n} c_k^i \left(\int_0^1 B_k^n(t)\mathcal{F}(t,\mathbf{y})\frac{d\Gamma}{d\xi}\frac{d\xi}{dt}\,dt\right) \end{aligned} \quad (21)$$

where, $\mathcal{F}(t,\mathbf{y})$ stands for the type of singularity in the fundamental solution.

This integral can be approximated using a quadrature rule of order $M$ as:

$$\begin{aligned} I &\simeq \sum_{k=0}^{n} c_k^i \left(\sum_{m=0}^{M} B_k^n(t_m)\mathcal{F}(t_m,\mathbf{y})\frac{d\xi}{dt}\frac{d\Gamma(t_m)}{d\xi}w_m\right) \\ &= \sum_{k=0}^{n} c_k^i \left(\sum_{m=0}^{M} \psi_k(t_m,\mathbf{y})\frac{d\Gamma(t_m)}{d\xi}w_m\right) \end{aligned} \quad (22)$$

where $t_m$ and $w_m$ are the integration points and weights, respectively. The quadrature weights are obtained from the solution of a system of equations defined from the above approximation [8]:

$$\sum_{m=0}^{M} \psi_k(t_m,\mathbf{y})w_m = m_k, \qquad k = 0,\ldots,n \quad (23)$$

where $m_k$ are the generalised moment:

$$m_k = \int_0^1 \psi_k(t_m,\mathbf{y})\,dt = \int_0^1 B_k^n(t)\mathcal{F}(t,\mathbf{y})\,dt \quad (24)$$

The generalised moments are the exact solution for the integral kernels in eqn (21) and can be calculated from the Brandaõ approach to the finite part integrals [10], [11].

The hypersingular integral kernels $\mathcal{F}(t,\mathbf{y})$, or its expansion in series, includes terms of the form $1/r^2$, with $r = ||\mathbf{x}-\mathbf{y}||$. The distance $r$ can be expressed in the univariate coordinate as $r = t_0 - 2t + 1$, where $t_0$ is the coordinate of the collocation point $\mathbf{y}$.

Then, the generalised moment of these terms are obtained according to the following formulas:

$$\begin{aligned} m_k &= \mathrm{FP}\int_0^1 \frac{B_k^n(t)}{(t_0-2t+1)^2}\frac{d\xi}{dt}\,dt \\ &= B_k^n(t_0)\mathrm{FP}\int_0^1 \frac{1}{(t_0-2t+1)^2}\frac{d\xi}{dt}\,dt - \frac{dB_k^n(t_0)}{dt}\frac{dt}{d\xi}\mathrm{CPV}\int_0^1 \frac{1}{t_0-2t+1}\frac{d\xi}{dt}\,dt \\ &+ \int_0^1 \frac{B_k^n(t) - B_k^n(t_0) + \dfrac{dB_k^n(t_0)}{dt}\dfrac{dt}{d\xi}(t_0-2t+1)}{t_0-2t+1}\frac{d\xi}{dt}\,dt \end{aligned} \quad (25)$$

where

$$\text{FP}\int_0^1 \frac{1}{(t_0-2t+1)^2}\frac{d\xi}{dt}\,dt = \begin{cases} 2/(t_0^2-1) & |t_0| \neq 1 \\ -1/2 & t_0 = \pm 1 \end{cases} \tag{26}$$

$$\text{CPV}\int_0^1 \frac{1}{t_0-2t+1}\frac{d\xi}{dt}\,dt = \begin{cases} \log\left|\dfrac{t_0+1}{1-t_0}\right| & |t_0| \neq 1 \\ \pm\log(2) & t_0 = \pm 1 \end{cases} \tag{27}$$

Eqn (23) defines a system of $n+1$ equations for each type of function to be integrated. The extension of the quadrature rule proposed in this work is capable of integrating kernels with constant, $\log(r)$, $1/r$, and $1/r^2$ terms when the collocation point belongs to the integration element, which are found in the fundamental solution or in their series expansions. Therefore, eqn (23) defines a system of $4(n+1)$ equations with $M+1$ unknown weights $w_m$. The solution is obtained in the least-squares sense when overdetermined and in the minimum norm least-squares sense when undetermined. The solution of the generalised moment for weakly singular and singular integrals can be found in Velázquez-Mata et al. [8].

### 2.4 Dealing with complex boundary geometry

The BEM formulation in the Bézier–Bernstein space [4], [9] is used in this work to handle exact geometries defined as Bézier curves, that are widely used in computer aid design (CAD). The element geometry is then represented as $\Gamma^j(\mathbf{x}) = \mathbf{r}_n^j(t)$, where the Bézier curves is defined as follows:

$$\mathbf{r}_n^j(t) = \sum_{k=0}^{n} \mathbf{b}_k^j B_k^n(t) \tag{28}$$

with $\mathbf{b}_k^j$ the control points used to approximate the geometry and $n$ the curve degree. Then, the element integral (eqn (18)) is rewritten in the univariate basis as [9]:

$$I = \int_{\Gamma^j} \phi^i(\xi)\mathcal{F}(\xi,\mathbf{y})\,d\Gamma = \int_0^1 \phi^i(\xi)\mathcal{F}(\xi,\mathbf{y})\left|\frac{d\mathbf{r}_n^j(t)}{dt}\right|dt \tag{29}$$

An efficient curve computation of $\mathbf{r}_n^j(t)$ is achieved using the polar form (or blossom) [12], which defines a multi-affine transformation satisfying:

$$\mathbf{b}_k^j = \mathbf{R}^j(\underbrace{0,\ldots,0}_{n-k},\underbrace{1,\ldots,1}_{k}) \tag{30}$$

where $\mathbf{R}^j(t_1,\ldots,t_n)$ is computed as:

$$\mathbf{R}^j(t_1,\ldots,t_n) = \sum_{\substack{I\cap K=\emptyset \\ I\cup K=\{1,2,\ldots,n\}}} \prod_{i\in I}(1-t_i)\prod_{k\in K} t_k \mathbf{b}_{|K|}^j \tag{31}$$

Thus, the Bézier in the polar form is defined as follows:

$$\mathbf{r}_n^j(t) = \sum_{k=0}^{n} \mathbf{R}^j(\underbrace{0,\ldots,0}_{n-k},\underbrace{1,\ldots,1}_{k})B_k^n(t) = \mathbf{R}^j(t,\ldots,t) \tag{32}$$

and its first derivative becomes:

$$\frac{d\mathbf{r}_n^j(t)}{dt} = \frac{d}{dt}\mathbf{R}^j(t,\ldots,t) = n\big(\mathbf{R}^j(\underbrace{t,\ldots,t}_{n-1},1) - \mathbf{R}^j(0,\underbrace{t,\ldots,t}_{n-1})\big) \tag{33}$$

Boundary elements are defined by subdividing Bézier curves according to a $h$-refinement strategy. The element geometry is evaluated and subdivided using the polar form of Bernstein polynomials without CAD interaction, providing great flexibility in mesh refinement. The field variables are approximated independently of the geometry representation, and therefore $p$-refinement only affects the field approximation.

### 2.5 System of equations

The system of equations is obtained from the discretised boundary integral equation according to the element approximation. Neumann and Robin boundary conditions are defined in scattering problems to define sound hard and impedance conditions from eqn (2):

$$\frac{\partial u(\mathbf{x})}{\partial \mathbf{n}(\mathbf{x})} = \frac{\gamma(\mathbf{x}) - \alpha(\mathbf{x})u(\mathbf{x})}{\beta(\mathbf{x})} \tag{34}$$

The boundary conditions can be implicitly defined in the integral kernels, rather than by prescribing nodal values as is done in standard formulations [4]. The main advantage of this procedure is that the right-hand side of the system of equations is integrated taking the exact distribution of loads into account. Then, eqn (6) becomes:

$$\begin{aligned} \frac{1}{2}u(\mathbf{y}) + \sum_{j=1}^{N}\left[\int_{\Gamma^j}\left(\frac{\alpha(\mathbf{x})}{\beta(\mathbf{x})}\Psi(\mathbf{x},\mathbf{y}) + \frac{\partial\Psi(\mathbf{x},\mathbf{y})}{\partial\mathbf{n}(\mathbf{x})}\right)u(\mathbf{x})\,d\Gamma\right] \\ = \sum_{j=1}^{N}\left[\int_{\Gamma^j}\frac{\gamma(\mathbf{x})}{\beta(\mathbf{x})}\Psi(\mathbf{x},\mathbf{y})\,d\Gamma\right] + u_0(\mathbf{y}) \end{aligned} \tag{35}$$

Once the element approximation (eqn (7)) is introduced in (35), a system of equations $\mathbf{H}\mathbf{u} = \mathbf{b}$ is obtained which allows the solution of the unknown nodal values $\mathbf{u}$. Similarly, a system of equation is $\widetilde{\mathbf{H}}\mathbf{u} = \widetilde{\mathbf{b}}$ is derived from the HBIE (eqn (12)):

$$\begin{aligned} -\frac{\alpha(\mathbf{y})}{2\beta(\mathbf{y})}u(\mathbf{y}) + \sum_{j=1}^{N}\left[\int_{\Gamma^j}\left(\frac{\alpha(\mathbf{x})}{\beta(\mathbf{x})}\frac{\partial\Psi(\mathbf{x},\mathbf{y})}{\partial\mathbf{n}(\mathbf{y})} + \frac{\partial^2\Psi(\mathbf{x},\mathbf{y})}{\partial\mathbf{n}(\mathbf{x})\partial\mathbf{n}(\mathbf{y})}\right)u(\mathbf{x})\,d\Gamma\right] \\ = -\frac{\gamma(\mathbf{y})}{2\beta(\mathbf{y})} + \sum_{j=1}^{N}\left[\int_{\Gamma^j}\frac{\gamma(\mathbf{x})}{\beta(\mathbf{x})}\frac{\partial\Psi(\mathbf{x},\mathbf{y})}{\partial\mathbf{n}(\mathbf{y})}\,d\Gamma\right] + \frac{\partial u_0(\mathbf{y})}{\partial\mathbf{n}(\mathbf{y})} \end{aligned} \tag{36}$$

Finally, the system of equation given by the Burton–Miller method using a linear combination of eqns (35) and (36) is $(\mathbf{H} + \mu\widetilde{\mathbf{H}})\mathbf{u} = \mathbf{b} + \mu\widetilde{\mathbf{b}}$, where $\mu$ is the coupling parameter.

The proposed method is valid for both thick and thin geometries. The boundary is represented by overlapping curves with outward normals in the limit case of zero thickness. The procedure for obtaining the system of equations and the solution is the same regardless of the boundary geometry. Therefore, the proposed BEM formulation is simplest and most general approach for dealing with arbitrary geometries.

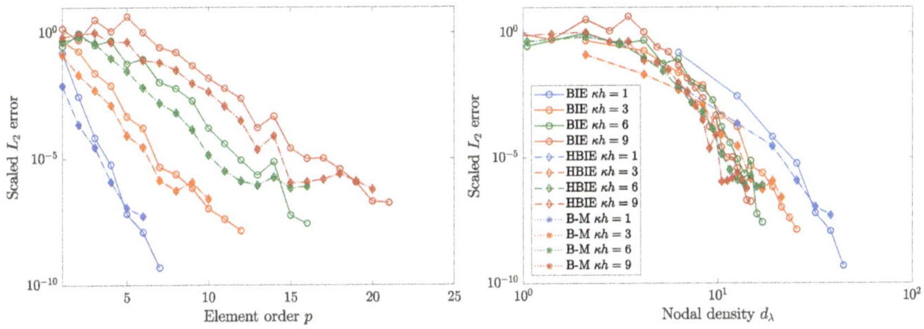


(a) (b)

Figure 2: Convergence analysis for $hp$-refinement plotted versus (a) element order $p$; and (b) nodal density per wavelength $d_\lambda$.

## 3 CHOOSING THE COUPLING PARAMETER

Many authors have discussed the choice of the coupling parameter $\mu$ as referred to by Marburg in [13]. The parameter $\mu = \pm\iota/\kappa$ is assumed to provide the system with the lowest condition number. Marburg [13] discussed that the correct sign of the coupling parameter depends on the harmonic time dependence and on the Green's function. Liu [3] suggests that using $\mu = \pm\iota h$ improves the contribution from both formulations, as the element length $h$ should decrease with increasing wavenumber. The author concludes that this choice improves the system conditioning and the stability of the results.

However, the choice of the coupling parameter for thin bodies has not been investigated as much in the past. The main contributions can be found in the early work of Rizzo and co-workers. Krishnasamy et al. [1] set $\mu = 1$ to study acoustic wave propagation through a thin penny-shaped scatterer with radius $a = 1$ and thickness up to $h = 10^{-7}$. Later, Liu and Rizzo [2] analysed shear waves in a penny-shaped open crack of different thicknesses and found that the choice of coupling parameter was not so restrictive. Values between $\mu = -1$ and $\mu = +1$ gave well-posed systems with low condition numbers.

This section study the optimum choice of the coupling parameter for thick and thin bodies.

### 3.1 Optimal choice of the coupling parameter for thick bodies

First, we considered a benchmark problem concerning the scattering of waves from a fixed cylindrical cavity at the origin with radius $r = 1\,\text{m}$ elicited by a source point in $\mathbf{y} = (0, 1)$. The circular geometry was approximated by a $C^2$ quartic Bézier curve [9]. The analytical solution to this problem can be found in Tadeu and Godinho [14].

Fig. 2 shows the $L_2$ scaled error $\epsilon_2$ for $\kappa = 40\,\text{rad/m}$ and different $hp-$strategies. A convergence analysis was carried out for several element sizes $h$ with successive $p-$enrichment. The element discretisation was set to $\kappa h = \{1,\ 3,\ 6,\ 9\}$, resulting in six, two, one and one-and-a-half elements per wavelength according to the node density:

$$d_\lambda = \frac{2\pi p}{\kappa h} \tag{37}$$

The boundary elements were then discretised to perform a $h-$ refinement and the element order was increased until convergence was reached. The coupling parameter was $\mu = -\iota h$ to account for the element length in the mesh refinements.

The lowest error was $\mathcal{O}(10^{-10})$ for the BIE formulation and $\mathcal{O}(10^{-8})$ for the HBIE and the Burton–Miller formulations. The accuracy of the Burton–Miller formulation was the same as that of the HBIE. Furthermore, the convergence was faster for the finer discretisation than for the coarser discretisation. The numerical results converged to the analytical solution when the node density per wavelength was higher than $d_\lambda = 3$ in the coarser meshes and $d_\lambda = 12$ for the finest discretisation. Furthermore, the solution error for a fixed node density was lower for a coarse mesh than for a fine mesh, as can be seen in Fig. 2(b). Therefore, a coarse mesh with an adequate node density gave better results than a fine mesh with the same density, or equivalently, the use of high order elements in coarse meshes ($p$-refinement) gave better results than low order elements in a fine discretisation ($h$-refinement) for a fixed node density.

We have found in several tests that the accuracy of the Burton–Miller method depends strongly on the choice of the coupling parameter. Although the most common choices such as $\mu = -\imath h$ or $\mu = \imath/\kappa$ give good results, there are other options that provide more accurate results. This issue requires further research to properly identify the optimal coupling parameter that minimises the solution error. The optimum choice of the coupling parameter was found as the optimal solution to this problem using a genetic algorithm. The polar form of the coupling parameter $\mu = \rho\angle\theta = \rho(\cos(\theta) + \imath\sin(\theta))$ was adopted for this analysis. The lower and upper limits of the optimisation variables were $\rho \in [0, 100]$ and $\theta \in [0, 2\pi]$. The initial population was 50, the crossover fraction was 0.8, and the elite children were 3 in each generation.

Fig. 3(a) compares the error for a wavenumber varying in the interval $\kappa \in [1,\ 80]$ rad/m obtained with the Burton–Miller method using the coupling parameters $\mu = -\imath/h$, $\mu = -\imath\kappa$ and the optimum choice $\mu = \rho\angle\theta$. The boundary discretisation was given by $\kappa h = 9$ with a node density $d_\lambda = 10$, which corresponds to one-and-a-half elements per wavelength of order $p = 15$. The error was $\mathcal{O}(10^{-6})$ for the common choices $\mu = -\imath/h$ and $\mu = -\imath\kappa$ of the coupling parameter. However, the error was $\mathcal{O}(10^{-10})$ for the optimal choice. Fig. 3(b) shows the radius and angle of the optimal coupling parameter. This fact requires further investigation.

Finally, Fig. 4 shows the scattered wavefield by degenerated ellipse with the minor axis tending to zero. The radial distribution of the incident wave field in the absence of the scatterer was lost under the effect of boundary wave reflections. A shadowed region was found downstream the scatterers in all cases where the potential amplitude was significantly lower.

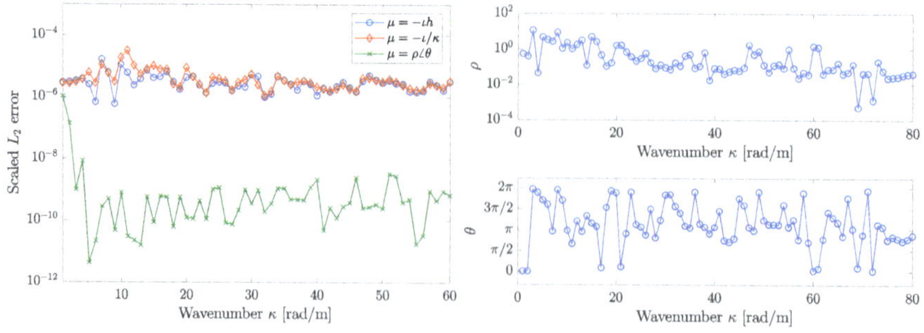


(a) (b)

Figure 3: (a) Scaled $L_2$ error for different choices of the coupling parameter; and (b) Optimal coupling parameter $\mu = \rho\angle\theta$.

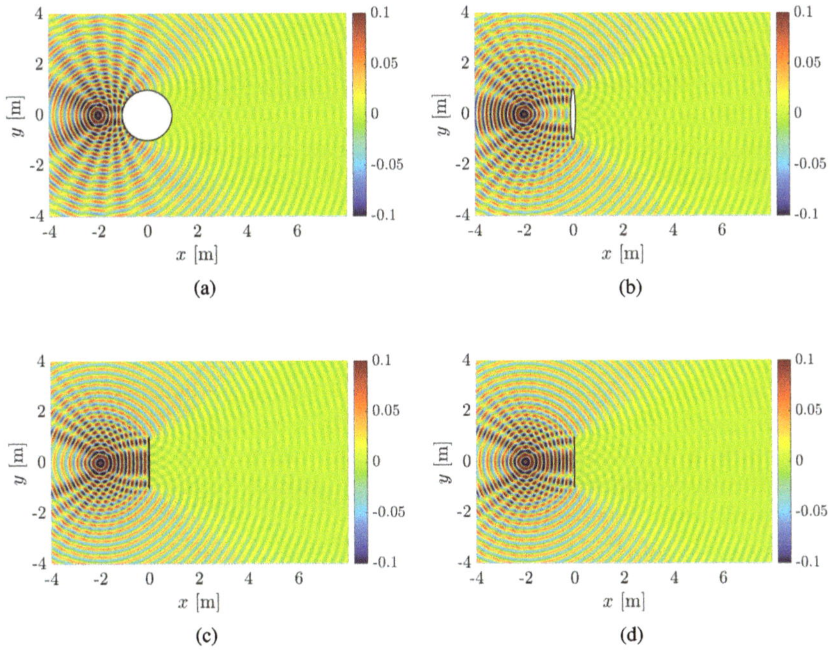


Figure 4: Real part of potential field distribution for ellipse geometries given by: (a) $R_x = 1.00\,\text{m}$; (b) $R_x = 0.10\,\text{m}$; (c) $R_x = 0.01\,\text{m}$; and (d) $R_x = 0.00\,\text{m}$.

## 4 CONCLUSIONS

This paper analysed the application of the Burton–Miller method to study wave scattering by thick or thin bodies with complex geometry. The proposed method is based on the Bézier–Bernstien formulation of the BEM. The approach allows the use of arbitrary high-order element and exact boundary geometry. The singular and hypersingular kernels are numerically evaluated without the need for a regularisation process. This approach is much simpler and more general than other dual formulations and improves the versatility of the BEM as both the BIE and HBIE are evaluated numerically. The procedure for obtaining the system of equations and the solution is the same regardless of the boundary geometry. Therefore, the proposed BEM formulation is the simplest and most general approach for dealing with arbitrary geometries.

## ACKNOWLEDGEMENTS

The authors would like to acknowledge the financial support provided by the Spanish Ministry of Science and Innovation under the research project PID2022-138674OB-C21, PROYEXCEL_00659 funded by Regional Ministry of Economic Transformation, Industry, Knowledge and Universities of Andalusia, and the Andalusian Scientific Computing Centre (CICA).

## REFERENCES

[1] Krishnasamy, G., Rizzo, F.J. & Liu, Y., Boundary integral equations for thin bodies. *International Journal for Numerical Methods in Engineering*, **37**(1), pp. 107–121, 1994.

[2] Liu, Y. & Rizzo, F., Scattering of elastic waves from thin shapes in three dimensions using the composite boundary integral equation formulation. *Journal of the Acoustical Society of America*, **102**(2 pt 1), pp. 926–932, 1997.
[3] Liu, Y., On the BEM for acoustic wave problems. *Engineering Analysis with Boundary Elements*, **107**, pp. 53–62, 2019.
[4] Romero, A., Galvín, P. & Tadeu, A., An accurate treatment of non-homogeneous boundary conditions for development of the BEM. *Engineering Analysis with Boundary Elements*, **116**, pp. 93–101, 2020.
[5] Wu, T., *Boundary Element Acoustics Fundamentals and Computer Codes. Advances in Boundary Elements*, WIT Press: Southampton and Boston, 2000.
[6] Toledo, R., Aznárez, J., Greiner, D. & Maeso, O., Shape design optimization of road acoustic barriers featuring top-edge devices by using genetic algorithms and boundary elements. *Engineering Analysis with Boundary Elements*, **63**, pp. 49–60, 2016.
[7] Kolm, P. & Rokhlin, V., Numerical quadratures for singular and hypersingular integrals. *Computers and Mathematics with Applications*, **41**(3), pp. 327–352, 2001.
[8] Velázquez-Mata, R., Romero, A., Domínguez, J., Tadeu, A. & Galvín, P., A novel high-performance quadrature rule for BEM formulations. *Engineering Analysis with Boundary Elements*, **140**, pp. 607–617, 2022.
[9] Romero, A., Galvín, P., Cámara-Molina, J. & Tadeu, A., On the formulation of a BEM in the Bézier–Bernstein space for the solution of Helmholtz equation. *Applied Mathematical Modelling*, **74**, pp. 301–319, 2019.
[10] Brandao, M.P., Improper integrals in theoretical aerodynamics: The problem revisited. *AIAA Journal*, **25**(9), pp. 1258–1260, 1987.
[11] Carley, M., Numerical quadratures for singular and hypersingular integrals in boundary element methods. *SIAM Journal on Scientific Computing*, **29**(3), pp. 1207–1216, 2007.
[12] Ramshaw, L., *Blossoming: A Connect-the-Dots Approach to Splines*. Digital Equipment Corporation SRC Report No. 19, 1987.
[13] Marburg, S., The Burton and Miller method: Unlocking another mystery of its coupling parameter. *Journal of Computational Acoustics*, **24**(1), 2016.
[14] Tadeu, A. & Godinho, L., Three-dimensional wave scattering by a fixed cylindrical inclusion submerged in a fluid medium. *Engineering Analysis with Boundary Elements*, **23**(9), pp. 745–755, 1999.
[15] Hur, S. & Kim, T.W., The best G1 cubic and G2 quartic Bézier approximations of circular arcs. *Journal of Computational and Applied Mathematics*, **236**(6), pp. 1183–1192, 2011.

# MULTISCALE MODELLING OF CONCRETE PLATES USING THE BOUNDARY ELEMENT METHOD

CALEB G. PITALUGA & GABRIELA R. FERNANDES
Civil Engineering Department, Federal University of Catalão (UFCAT), Brazil

## ABSTRACT

In this paper we analyse a concrete plate by using a multi-scale modelling technique. The boundary element nonlinear formulation for the two-dimensional problem is used to model the macrocontinuum while the concrete microstructure is modelled by a boundary element nonlinear formulation based on the concept of representative volume element (RVE) and considering it as a zoned plate where each sub-region represents a RVE phase. In both formulations the consistent tangent operator is used to achieve the convergence of the iterative procedures. The equilibrium problem of the plate is solved in terms of in-plane strains while for the RVE equilibrium problem it is done in terms of displacement fluctuations. To model the concrete microstructure, the Mohr–Coulomb criterion is used to govern the mechanical behaviour of the mortar matrix, while the aggregates are considered elastics. Besides, some voids are also defined in the mortar matrix to model the concrete porosity and the fracture process along the interfaces between matrix and aggregates are modelled by defining addition cohesive-contact finite elements superposed to the interface elements. To validate the presented model, the numerical results are compared to experimental ones.

*Keywords: multi-scale modelling, homogenisation, RVE, boundary elements, 2D problem.*

## 1 INTRODUCTION

In the present model, the concrete microstructure is numerically modelled by the representative volume element (RVE) formulation, and the RVE homogenised response defines the concrete constitutive model. For that, the strain tensor related to the macrocontinuum point must be imposed to the RVE and its equilibrium problem solved (see [1]–[6]). Therefore, to solve the multiscale problem of the plate, we must assign a RVE for each point considered in the plate equilibrium equation. Thus, in the present model two different formulations must be coupled to obtain the solution, both developed with the boundary element method (BEM): the formulation of the macrocontinuum (plate) (see Fernandes and de Souza Neto [7]) and the RVE formulation (concrete microstructure). As we deal with dissipative phenomena in the microstructure, both formulations require an iterative procedure to achieve the equilibrium of its respective problem. Thus, for each iteration of the macrocontinuum problem, we must solve an iterative procedure at each macrocontinuum point, where the RVE is assigned. Therefore, the multi-scale modelling is very expensive computationally, what encourages us to investigate other numerical models that reduce this computational effort, as the ones developed by the BEM.

In this study, to model the concrete microstructure, we consider the aggregates as elastic inclusions defined inside a mortar matrix, where some voids are also defined to model the concrete porosity. The Mohr–Coulomb criterion is adopted to govern the mechanical behaviour of the mortar (see Silva et al. [8]) and we define additional cohesive-contact finite elements on some interfaces to model the phase debonding phenomenon (see Pituba et al. [9]). These finite elements are defined superposed to BEM interface elements, being their mechanical behaviour governed by either a contact model or cohesive fracture model.

In the numerical example of a concrete plate subjected to compression load, we show that the present model can reproduce the experimental results (obtained in Delalibera [10]), showing its efficiency and accuracy.

WIT Transactions on Engineering Sciences, Vol 136, 
www.witpress.com, ISSN 1743-3533 (on-line)
doi:10.2495/BE470121

## 2 MACRO-SCALE: THE NON-LINEAR TWO-DIMENSIONAL PROBLEM

Let us consider a plate of thickness $t$, represented by its middle surface defined in the plane ($x_1$, $x_2$), where the following values are defined in terms of normal and tangential directions: in-plane tractions ($\dot{p}_n$ and $\dot{p}_s$) and in-plane displacements ($\dot{u}_n$ and $\dot{u}_s$). As we deal with plasticity and fracture process at the material microstructure, all variables are expressed in their time derivatives, i.e., $(\dot{x}) = dx/dt$ or in terms of their increments ($\Delta x$). In this context, the total strain is split into its elastic ($\dot{\varepsilon}_{ij}^e$) and inelastic ($\dot{\varepsilon}_{ij}^0$) parts and the inelastic membrane force rate $\dot{N}_{ij}^0$, is defined as:

$$\dot{N}_{ij}^0 = \dot{N}_{ij}^e - \dot{N}_{ij} \qquad i, j = 1, 2, \tag{1}$$

where the forces $\dot{N}_{ij}$ satisfy the constitutive model, and the forces $\dot{N}_{ij}^e$ are defined using the total strain $\dot{\varepsilon}_{ij}$ and the Hooke's law. For plane stress condition, they are given by:

$$\dot{N}_{ij}^e = \frac{\bar{E}}{(1-\nu^2)}\left[\nu \;\; \dot{\varepsilon}_{kk}\delta_{ij} + (1-\nu \;\;)\dot{\varepsilon}_{ij}\right] \qquad i, j, k = 1, 2, \tag{2}$$

where $\delta_{ij}$ is the Kronecker delta, $\bar{E} = Et$ being $E$ the Young's modulus, $\nu$ the Poisson's ration. The integral representation for in-plane displacements can be obtained by applying the Betti's reciprocal theorem written in terms of forces $\dot{N}_{ij}^e$. After integrating by parts and considering eqn (1), we obtain the following representation of in-plane displacements:

$$K_i(q)\dot{u}_i(q) = -\int_\Gamma (\dot{u}_n p_{in}^* + \dot{u}_s p_{is}^*)d\Gamma + \int_\Gamma (u_{in}^* \dot{p}_n + u_{is}^* \dot{p}_s)d\Gamma + \int_{\Omega_b} (u_{in}^* \dot{b}_n + u_{is}^* \dot{b}_s)d\Omega$$

$$+ \int_\Omega \varepsilon_{ijk}^* \dot{N}_{jk}^0 \, d\Omega \qquad k, i, j = 1,2, \tag{3}$$

where $q$ is the source point, the superscript * denotes the fundamental values; $i$ denotes the fundamental load direction; $\Omega_b$ is the plate loaded area; the free term $K_i(q)$ is defined as $K_i(q) = 1$ and $K_i(q) = 1/2$, respectively, for internal and boundary points not coincident to corners. The integral representation for rotations is obtained by differentiating eqn (3):

$$\Delta u_{i,\ell}(q) = -\int_\Gamma (\Delta u_n p_{in,\ell}^* + \Delta u_s p_{is,\ell}^*)d\Gamma + \int_\Gamma (u_{in,\ell}^* \Delta p_n + u_{is,\ell}^* \Delta p_s)d\Gamma$$

$$+ \int_{\Omega_b} (u_{in,\ell}^* \Delta b_n + u_{is,\ell}^* \Delta b_s)d\Omega + \int_\Omega \frac{\partial \varepsilon_{ijk}^*}{\partial x_l} \Delta N_{jk}^0 \, d\Omega$$

$$+ \frac{1}{16\bar{G}(1-\nu')}\left[(6 - 8\nu')\dot{N}_{il}^0(q) - \dot{N}_{kk}^0(q)\delta_{il}\right] \qquad k, j, l = 1,2. \tag{4}$$

Now we must transform the integral representations (eqns (3) and (4)) into algebraic equations. For that, we consider linear elements where the displacements and tractions are approximated by quadratic functions. The elements are not discontinuous, i.e., their initial and final nodes correspond to the element extremities. At corners we adopt duplicate nodes and to write the algebraic equation for these nodes we move the node inside the element. Besides, we have also to discretise the plate domain because of the inelastic forces defined in the domain. For that we adopt triangular cells where the inelastic forces are approximated by linear functions. To write the set of equations necessary to obtain the boundary values, we consider the two in-plane displacements equation written in each boundary node. After applying the boundary conditions to the boundary nodes, the unknown's vector on the plate boundary ($\boldsymbol{\Delta X}$) are computed from the following set of equations:

$$\boldsymbol{\Delta X} = \boldsymbol{\Delta L} + \boldsymbol{R}_N \boldsymbol{\Delta N}^0, \tag{5}$$

where vector $\boldsymbol{\Delta L}$ defines the elastic solution and $\boldsymbol{R}_N$ are the corrections due to the dissipative forces ($\boldsymbol{\Delta N}^0$).

In eqn (5) $\boldsymbol{R}_N = \boldsymbol{A}^{-1}\boldsymbol{E}$, where the matrix $\boldsymbol{E}$ represents the integration over the cells; the matrix $\boldsymbol{A}$ is a mix of the matrices $\boldsymbol{H}$ and $\boldsymbol{G}$ usually defined in the boundary element method. The integral representation for the elastic forces $\dot{N}_{ij}^{e}$ can be obtained from eqn (4), after applying the Hooke's law. After transforming eqn (4) into an algebraic equation, we can write the following algebraic equation of elastic force increments at each cell node:

$$\boldsymbol{\Delta N}^{e(BEM)} = \boldsymbol{\Delta K}_N + \boldsymbol{S}_N \boldsymbol{\Delta N}^0, \tag{6}$$

where $\boldsymbol{\Delta K}_N$ is the elastic solution, and $\boldsymbol{S}_N$ defines the corrections due to the dissipative forces $\boldsymbol{\Delta N}^0$.

To define the plate equilibrium equation, we need now to write algebraic equation for the forces $\boldsymbol{\Delta N}^{BEM}$ computed considering the stress field that satisfy the constitutive model (for more details see Fernandes and de Souza Neto [7]):

$$\boldsymbol{\Delta N}^{BEM} = \boldsymbol{C}_N \boldsymbol{\Delta \varepsilon} - \boldsymbol{\Delta K}_N - \boldsymbol{S}_N \boldsymbol{\Delta N}^0 + \boldsymbol{\Delta N}, \tag{7}$$

where the $\boldsymbol{C}_N$ matrix defines the elastic matrix of all cell nodes; for a node $k$, $\boldsymbol{\Delta N}^{0(k)}$ is computed considering its total strain $\boldsymbol{\Delta \varepsilon}$ and forces $\boldsymbol{\Delta N}$, i.e., $\boldsymbol{\Delta N}^{0(k)} = \boldsymbol{C}_N^{(k)} \boldsymbol{\Delta \varepsilon}^{(k)} - \boldsymbol{\Delta N}^{(k)}$, where $\boldsymbol{\Delta N}^{(k)}$ satisfies the constitutive model or in our case, it is computed from the homogenised stress vector of the RVE.

Considering a load increment $n$, the plate equilibrium is achieved, when the following expression is satisfied:

$$\boldsymbol{\Delta K}_{N(n)} - \boldsymbol{\Delta N}^{BEM} = \boldsymbol{0}. \tag{8}$$

Replacing eqn (7) into eqn (8) the plate equilibrium equation can be expressed in terms of total in-plane strains:

$$\boldsymbol{R}_N(\boldsymbol{\Delta \varepsilon}_n) = 2\boldsymbol{\Delta K}_{N(n)} - \boldsymbol{C}_N \boldsymbol{\Delta \varepsilon}_n + \boldsymbol{S}_N(\boldsymbol{C}_N \boldsymbol{\Delta \varepsilon}_n - \boldsymbol{\Delta N}_n) - \boldsymbol{\Delta N}_n = \boldsymbol{0}. \tag{9}$$

When all plate points behave elastically eqn (9) is automatically satisfied. Otherwise an iterative procedure ($i \geq 1$) is required to obtain residual forces ($\boldsymbol{R}_N$) null, where additive corrections $\boldsymbol{\delta\Delta\varepsilon}_n^{i+1}$ must be added to the macro-strain vector ($\boldsymbol{\Delta\varepsilon}_n^i$), i.e., $\boldsymbol{\Delta\varepsilon}_n^{i+1} = \boldsymbol{\Delta\varepsilon}_n^i + \boldsymbol{\delta\Delta\varepsilon}_n^{i+1}$. For that, we apply the Newton–Raphson's scheme to eqn (9). Initially, the strain increment $\boldsymbol{\Delta\varepsilon}_n^0$ is computed from the elastic solution ($\boldsymbol{\Delta K}_N$) for all cell nodes and a RVE is assigned to each one of these nodes. After imposing $\boldsymbol{\Delta\varepsilon}_n^0$ to its respective RVE, we obtain the RVE homogenised response for all cell nodes and eqn (9) is checked. We consider that eqn (9) is satisfied when $\frac{\sqrt{\boldsymbol{R}_N^T \boldsymbol{R}_N}}{\sqrt{\boldsymbol{\Delta K}_N^T \boldsymbol{\Delta K}_N}} \leq tol$, *tol* being a given tolerance. If it is not satisfied, new corrections ($\boldsymbol{\delta\Delta\varepsilon}_n^{i+1}$) must be obtained and the process continues at iteration i + 1. The corrections $\boldsymbol{\delta\Delta\varepsilon}_n^{i+1}$ are obtained from the linearisation of eqn (9) i.e.,

$$\delta\Delta\varepsilon_n^{i+1} = -\left[\frac{\partial R_N(\Delta\varepsilon_n^i)}{\partial\Delta\varepsilon_n^i}\right]^{-1} R_N(\Delta\varepsilon_n^i), \tag{10}$$

where: $-\frac{\partial R_N\left(\Delta\varepsilon_n^{(i)}\right)}{\partial\Delta\varepsilon_n^{(i)}} = S_N\left(C_{N(n)}^{ep(i)} - C_N\right) + C_N + C_{N(n)}^{ep(i)}$ is the consistent tangent operator (CTO), obtained by differentiating eqn (9). The $C_{N(n)}^{ep(i)}$ matrix contains the $C_N^{ep}$ tensor of all cell points. For a node *k*, $C_N^{ep(k)} = tC^{ep(k)}$, where $C^{ep(k)}$ is the inelastic tangent constitutive tensor of the cell node *k* (it is given by the homogenised constitutive tensor of the RVE assigned for node *k*).

## 3 BEM FORMULATION FOR THE MICROCONTINUUM PROBLEM

In the multi-scale analysis, the constitutive model is defined by the RVE homogenised response. Therefore, one RVE must be assigned for each cell node ($x$) of the macro-continuum (see Fig. 1, where $y$ denotes a RVE point and $\Omega_\mu$ the RVE domain). The RVE has a matrix $\Omega_l$ where we can insert inclusions and/or voids (see Fig. 2). Besides, along the interfaces between matrix and inclusions we can define cohesive-contact finite elements to model the phase debonding that occurs during the fracture process (see Pituba et al. [9]). If these cohesive-contact elements are not defined, the interface is considered perfectly bonded.

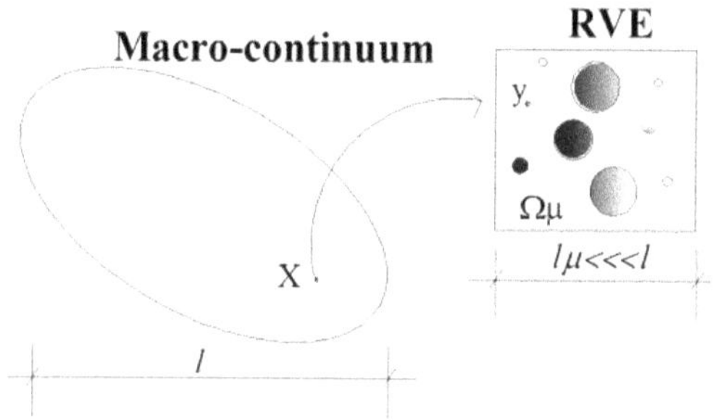


Figure 1: Representation of the macro-continuum and the RVE.

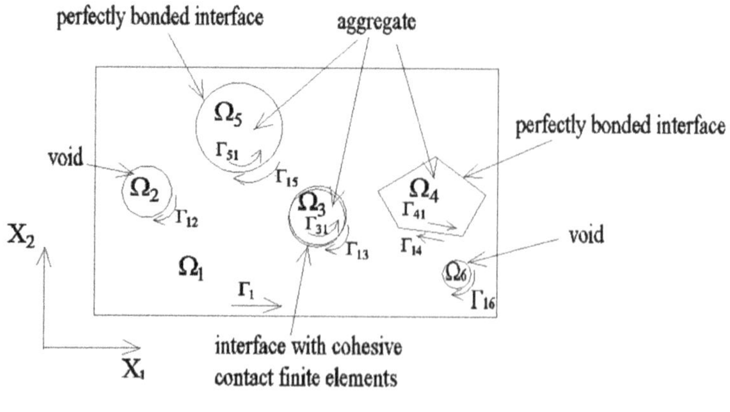


Figure 2: RVE modelling a heterogeneous microstructure where the microcracking process can occur in the ITZ region.

The integral representation of in-plane displacements ($\dot{u}_i^\mu$) in the microstructure can be obtained from Bettis's theorem, considering a zoned plate where each sub-region can have different Young's modulus and Poisson's ratio. It is also important to stress that the material behaviour at each RVE phase (matrix and inclusions) can be governed by different constitutive models. The integral representation of in-plane displacements ($\dot{u}_i^\mu$) in the RVE is defined as (see Fernandes et al. [5], [6] and Silva et al. [8]):

$$C_{k1}\dot{u}_1^\mu(s) + C_{k2}\dot{u}_2^\mu(s) = -\frac{\overline{E}_1\nu_1}{\overline{E}\nu}\int_{\Gamma_1}\left(\dot{u}_1^\mu(P)p_{k1}^{*(\mu)}(s,P) + \dot{u}_2^\mu(P)p_{k2}^{*(\mu)}(s,P)\right)d\Gamma +$$

$$-\sum_{m=1}^{N_{inc}}\frac{(\overline{E}_m\nu_m - \overline{E}_1\nu_1)}{\overline{E}\nu}\int_{\Gamma_{m1}}\left(\dot{u}_1^\mu(P)p_{k1}^{*(\mu)}(s,P) + \dot{u}_2^\mu(P)p_{k2}^{*(\mu)}(s,P)\right)d\Gamma_{m1}$$

$$-\sum_{m=1}^{N_{voids}}\frac{\overline{E}_1\nu_1}{\overline{E}\nu}\int_{\Gamma_{1m}}\left(\dot{u}_1^\mu(P)p_{k1}^{*(\mu)}(s,P) + \dot{u}_2^\mu(P)p_{k2}^{*(\mu)}(s,P)\right)d\Gamma_{1m}$$

$$+\int_{\Gamma_1}\left(u_{k1}^{*(\mu)}(s,P)\dot{p}_1^\mu(P) + u_{k2}^{*(\mu)}(s,P)\dot{p}_2^\mu(P)\right)d\Gamma$$

$$-\overline{E}_1\left(1-\frac{\nu_1}{\nu}\right)\int_{\Gamma_1}\left[\dot{u}_1^\mu(P)\varepsilon_{k1}^{*(\mu)}(s,P) + \dot{u}_2^\mu(P)\varepsilon_{k2}^{*(\mu)}(s,P)\right]d\Gamma +$$

$$-\sum_{m=1}^{N_{inc}}\left[\overline{E}_m\left(1-\frac{\nu_m}{\nu}\right) - \overline{E}_1\left(1-\frac{\nu_1}{\nu}\right)\right]\int_{\Gamma_{m1}}\left[\dot{u}_1^\mu(P)\varepsilon_{k1}^{*(\mu)}(s,P) + \dot{u}_2^\mu(P)\varepsilon_{k2}^{*(\mu)}(s,P)\right]d\Gamma_{m1}$$

$$+\sum_{m=1}^{N_s}\overline{E}_m\left(1-\frac{\nu_m}{\nu}\right)\int_{\Omega_m}\dot{u}_i^\mu(p)\varepsilon_{kij,j}^{*(\mu)}(s,p)d\Omega_m + \int_\Omega \varepsilon_{kij}^{*(\mu)}\dot{N}_{ij}^{0(\mu)}\,d\Omega \qquad k,i,j = 1,2, \tag{11}$$

where the fundamental values (terms with *), $\nu$ and $\overline{E}$ are related to the sub-region where the source point $s$ is placed, $\overline{E}_m = \frac{\bar{E}_m}{(1-\nu_m{}^2)}$, $\bar{E}_m = E_m t$; $E$ is Young's modulus, $\nu$ the Poisson's ratio, $t$ the plate thickness and $\delta_{ij}$ the Kronecker's delta; $N_s$ the sub-regions number; $k$ the fundamental load direction; the subscript '1' refers to the matrix; $\Gamma_1$ is the external boundary; $\Gamma_{m1}$ represents a matrix/inclusion interface; $\Gamma_{1m}$ is a matrix/void interface. For internal collocation points, $C_{ki} = 1$ and $C_{ki} = 0$, respectively, for $k = i$ and $k \neq i$; for boundary or interface collocation points, the free terms $C_{k1}$ and $C_{k2}$ can be found in the works of Fernandes et al. [4], [5]. All fundamental expressions can be found in Fernandes et al. [4].

In eqn (11), $\dot{N}_{ij}^0$ are dissipative forces, which are defined as:

$$\dot{N}_{ij}^{0(\mu)} = \dot{N}_{ij}^{p(\mu)} - \dot{N}_{ij}^{cf(\mu)}, \tag{12}$$

where $\dot{N}_{ij}^{p(\mu)}$ is related to the plastic process and $\dot{N}_{ij}^{cf(\mu)}$ to the phase debonding; these forces are defined as:

$$\dot{N}_{ij}^{p(\mu)} = \dot{N}_{ij}^{e(\mu)} - \dot{N}_{ij}^{\mu}, \tag{13}$$

$$\dot{N}_{ij}^{cf(\mu)} = \dot{N}_{ij}^{e(\mu)} - \dot{\tilde{N}}_{ij}^{\mu}. \tag{14}$$

In eqn (14) the continuous forces $\dot{\tilde{N}}_{ij}^{\mu}$ are given by the elastic forces computed before solving the RVE equilibrium problem (see Fernandes et al. [5]). Note that after achieving the RVE equilibrium, the strain field becomes discontinuous if the phase debonding occurred on interfaces. In eqn (13) the forces $\dot{N}_{ij}^{\mu}$ satisfy the constitutive model adopted to govern mechanical behaviour in the RVE phase, while the elastic membrane forces $\dot{N}_{ij}^{e(\mu)}$ are computed from the total strain, i.e.,

$$\dot{N}_{ij}^{e(\mu)} = \frac{\bar{E}}{(1-\nu^2)}\left[\nu\,\dot{\varepsilon}_{kk}^{\mu}\delta_{ij} + (1-\nu)\dot{\varepsilon}_{ij}^{\mu}\right]. \tag{15}$$

To compute the unknown's problem, we must transform eqn (11) into its algebraic form by considering linear elements on the boundary and interfaces and triangular cells in the domain. In the elements the displacements and tractions vary linearly while in the cells the displacements are approximated by linear functions and the dissipative forces are adopted constant. In the RVE there are two unknowns at boundary nodes (displacements or tractions, according to boundary conditions) and two unknowns at interface and cell nodes (displacements). Thus, to obtain the set of equations necessary to solve the problem, we write two displacement equations at boundary, interface, and cell nodes, resulting into:

$$\begin{bmatrix} [H]_{BB} & [H]_{Bi} \\ [H]_{iB} & [H]_{ii} \end{bmatrix}_{RVE} \begin{Bmatrix} \{\Delta U\}_B \\ \{\Delta U\}_i \end{Bmatrix}_{RVE} = \begin{bmatrix} [G]_{BB} \\ [G]_{iB} \end{bmatrix}_{RVE} \{\Delta P_B\}_{RVE} + \begin{bmatrix} [E]_{BB} \\ [E]_{iB} \end{bmatrix}_{RVE} \{\Delta N^0\}_{RVE}, \tag{16}$$

where ***U*** and ***P*** refer, respectively to nodal displacements and tractions, subscript *B* refers to the boundary, subscript *i* refers to interface or internal node; in a simplified form eqn (16) is defined as:

$$[H]_{RVE}\{\Delta U\}_{RVE} = [G]_{RVE}\{\Delta P_B\}_{RVE} + [E]_{RVE}\{\Delta N^0\}_{RVE}. \tag{17}$$

In the RVE the boundary conditions are defined by the imposition of the strain vector of the macro-continuum point, i.e., a constant strain is imposed to the RVE what implies to impose linear displacements to the boundary nodes. After imposing these boundary conditions to eqn (17), the unknown values ($\Delta X_{RVE}$) can be computed (see Fernandes et al. [4]):

$$\{\Delta X\}_{RVE} = \{\Delta L\}_{RVE} + [R]_{RVE}\{\Delta N^0\}_{RVE}, \tag{18}$$

where $\{\Delta X\}_{RVE} = \begin{Bmatrix} \{\Delta P\}_B \\ \{\Delta U\}_i \end{Bmatrix}_{RVE}$, $\{\Delta L\}_{RVE}$ represents the elastic solution; $\{\Delta L\}_{RVE} = [A]_{RVE}^{-1}\{\Delta B\}_{RVE}$; $[R]_{RVE} = [A]_{RVE}^{-1}[E]_{RVE}$; $\{\Delta B\}_{RVE} = \begin{bmatrix} -[H]_{BB} \\ -[H]_{iB} \end{bmatrix}_{RVE} \{\Delta U_B\}_{RVE}$; $[A]_{RVE} = \begin{bmatrix} -[G]_{BB} & [H]_{Bi} \\ -[G]_{iB} & [H]_{ii} \end{bmatrix}_{RVE}$.

Solving the problem, however, requires the algebraic equation of the membrane forces, $\{\Delta\bar{N}\}_{RVE}^{BEM}$ at the centre of each cell, defined by:

$$\{\Delta\bar{N}\}^{BEM} = -[[H']_B \quad [H']_i]_{RVE} \begin{Bmatrix} \{\Delta U\}_B \\ \{\Delta U\}_i \end{Bmatrix}_{RVE} + [G_B']_{RVE}\{\Delta P_B\}_{RVE} + [E']_{RVE}\{\Delta N^0\}_{RVE}. \tag{19}$$

To solve eqn (19) we have to impose linear displacements to the boundary nodes, resulting in:

$$\{\Delta\bar{N}\}_{RVE}^{BEM} = \{\Delta K\}_{RVE} + [S]_{RVE}\{\Delta N^0\}_{RVE}. \tag{20}$$

where$\{\Delta K\}_{RVE} = \{\Delta B'\}_{RVE} - [A']_{RVE}\{\Delta L\}_{RVE}$;$[S]_{RVE} = [E']_{RVE} - [A']_{RVE}[R]_{RVE}$; $\{\Delta B'\}_{RVE} = -[H_B']_{RVE}\{\Delta U_B\}_{RVE}$; $[A']_{RVE} = [-[G_B']_{RVE} \quad [H_i']_{RVE}]$.

On the other hand, the RVE displacement field, $\boldsymbol{u}_\mu(\boldsymbol{y})$, contains the displacement field due to the imposed macroscopic strain ($\boldsymbol{u}_\mu^\varepsilon(\boldsymbol{y})$) and the displacement fluctuation field ($\tilde{\boldsymbol{u}}_\mu(\boldsymbol{y})$), i.e.,

$$\boldsymbol{u}_\mu(\boldsymbol{y}) = \boldsymbol{u}_\mu^\varepsilon(\boldsymbol{y}) + \tilde{\boldsymbol{u}}_\mu(\boldsymbol{y}). \tag{21}$$

Accordingly, the microscopic strain field can also be decomposed into the macroscopic strain $\boldsymbol{\varepsilon}(\boldsymbol{x}) = \boldsymbol{\varepsilon}$ and the strain fluctuation field $\tilde{\boldsymbol{\varepsilon}}_\mu$ ($\tilde{\boldsymbol{\varepsilon}}_\mu = \boldsymbol{B}\tilde{\boldsymbol{u}}_\mu$), where $\boldsymbol{B}$ is the strain-displacement matrix of the finite element method).

In turn, the RVE equilibrium problem is achieved when the stress field is self-equilibrated. For that, the following equilibrium equation must be satisfied:

$$R_F = \sum_{e=1}^{N_{cell}^\mu} B_e^T \Delta\sigma_\mu^{e(i)} t\, A_e = \sum_{e=1}^{N_{cell}^\mu} [B]_e^T t [D_\mu^e]\left(\{\Delta\varepsilon\} + [B]\{\Delta\tilde{u}_\mu\}\right)_e A_e = 0, \tag{22}$$

where subscript *e* refers to a cell; $N_{cell}^\mu$ is the cell number in the RVE domain; *t* is the macro-continuum thickness; $[D_\mu^e]$ is the constitutive tensor, $B_e$ the strain-displacement matrix; $A_e$ the cell area.

If eqn (22) is not verified an iterative procedure $i_{RVE} \geq 0$ is required where corrections $\delta\tilde{u}_\mu^{i_{RVE}+1}$ must be added to the fluctuation increment at iteration $i_{RVE}+1$, i.e., $\Delta\tilde{u}_\mu^{i_{RVE}+1} = \Delta\tilde{u}_\mu^{i_{RVE}} + \delta\tilde{u}_\mu^{i_{RVE}+1}$. These corrections are computed from the following expression:

$$\Delta\boldsymbol{F}^{i_{RVE}} + \boldsymbol{K}^{i_{RVE}}\delta\tilde{\boldsymbol{u}}_\mu^{i_{RVE}+1} = 0, \tag{23}$$

where the matrix *K* is the CTO operator and F the nodal forces vector given by:

$$\boldsymbol{K}^{i_{RVE}} = \sum_{e=1}^{N_{cel}} \boldsymbol{B}_e^T \boldsymbol{D}_\mu^{(e)i_{RVE}} \boldsymbol{B}_e t\, A_e + \sum_{ef=1}^{N_f} \boldsymbol{K}_{ef}^{i_{RVE}}\, t, \tag{24}$$

$$\Delta\boldsymbol{F}^{i_{RVE}} = \sum_{e=1}^{N_{cel}} \boldsymbol{B}_e^T \Delta\boldsymbol{\sigma}_\mu^{e(i_{RVE})} t\, A_e + \sum_{ef=1}^{N_f} \Delta\boldsymbol{F}_{ef}^{i_{RVE}}\, t, \tag{25}$$

being $N_f$ the number of cohesive contact fracture element; subscript *e* and *ef* refers, respectively to the triangular cell and cohesive contact element.

These cohesive contact finite elements are rectangular; they have four nodes and are placed on the interfaces between inclusions and matrix to model the phase debonding (see Fig. 3). Their nodes are adopted coincident to the interface and cell nodes (see Fig. 3). For more details about the definition of $\boldsymbol{K}_{ef}$ and $\boldsymbol{F}_{ef}$, please see Pituba et al. [9].

The RVE homogenised constitutive tensor is defined as:

$$\boldsymbol{C}^{ep} = \boldsymbol{C}^{ep(Taylor)} + \tilde{\boldsymbol{C}}^{ep}, \tag{26}$$

where $\boldsymbol{C}^{ep(Taylor)} = \sum_{p=1}^{N_p} \frac{V_p}{V_\mu} \boldsymbol{D}_\mu^{(p)}$;$\tilde{\boldsymbol{C}}^{ep} = -\frac{1}{V_\mu} \boldsymbol{G}_R \boldsymbol{K}_R^{-1} \boldsymbol{G}_R^T$; $N_p$ is the number of phases of the RVE; $V_\mu$ is the RVE volume; $\boldsymbol{G} = \sum_{e=1}^{N_{cell}^{RVE}} \boldsymbol{D}_\mu^e \boldsymbol{B}_e \boldsymbol{V}_e$.

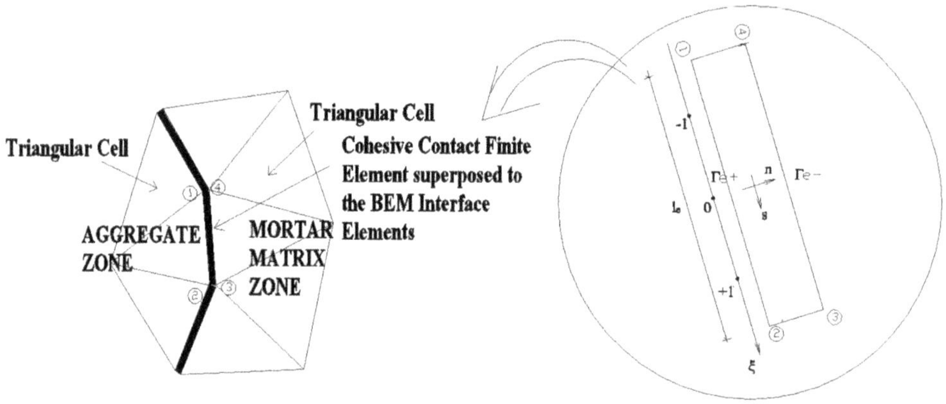


Figure 3: Cohesive contact finite element and its insertion between triangular cells.

The $C^{ep(Taylor)}$ tensor is obtained by assuming the displacement fluctuation field null while $\tilde{C}^{ep}$ represents the influence of the displacement fluctuation. The RVE homogenised stress tensor is defined as:

$$\boldsymbol{\sigma} = (\bar{\boldsymbol{\sigma}} + \bar{\boldsymbol{\sigma}}^T)\frac{1}{2V_\mu}, \tag{27}$$

where $\bar{\sigma} = \sum_{i=1}^{N_b}\{F\}_i\{y\}_i^T$, $N_b$ being the number of nodes on the external boundary, $\boldsymbol{F}$ the forces at these boundary nodes and $\boldsymbol{y}$ their coordinates vector. The forces $\boldsymbol{F}$ are defined as:$\{F_i\}_n = \{P_B\}_n\frac{1}{2}(L_{e-1} + L_e)$, where $\{P_B\}_n = \{P_B\}_{n-1} + \{\Delta P_B\}_n$, $L$ is the element length, $n$ is the increment of the macro-continuum problem; $\{\Delta P_B\}_n$ is obtained from eqn (16).

Note that when the RVE equilibrium is achieved, we must add the displacement increment fluctuation field to the nodal boundary displacements $\{\Delta U_B\}_n$. Then, we must update the boundary tractions $\{\Delta P_B\}_n$ by using eqn (16).

## 4 NUMERICAL EXAMPLE

Consider the plate depicted in Fig. 4(a), subjected to compression force. Its discretisation is depicted in Fig. 4(b), where only one quarter of the plate is defined, for symmetry reasons. The plate thickness is 1mm and it is discretised by 16 boundary elements and 16 cells. We define two elements at each smaller side and six elements along each bigger side. The plate boundary conditions are defined as: tractions $p_s = 0$ $p_n = -25$ N/mm to the nodes 19 to 23; tractions $p_s = p_n = 0$ to nodes 6 to 18; $u_n = 0$ and $p_s$=0 to the nodes 1 to 5 and to the nodes 24 to 36.

The concrete microstructure is modelled considering two RVEs (see Fig. 5). RVE-25i is composed of 25 aggregates of different dimensions (see Table 1) and 12 voids of 6 mm diameter, randomly distributed in the mortar domain; RVE-5i is defined by five aggregates (four of 33 mm diameter and one of 27 mm diameter) and four voids of 10.4 mm diameter, symmetrically distributed in the domain. The volume fractions of inclusions and voids remains the same, being defined, respectively, as 40% and 3.39%. Their discretisation is defined in Table 2.

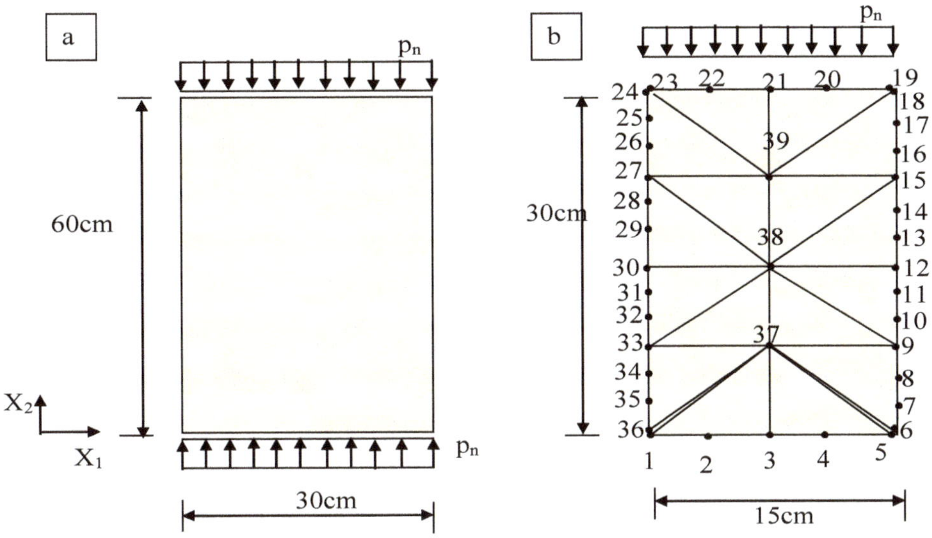


Figure 4: (a) Geometry and loads of the plate; and (b) Discretisation of the plate.

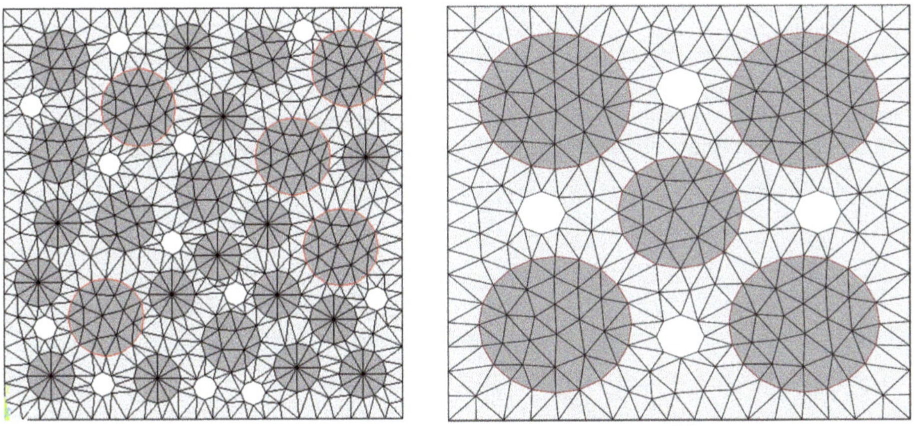

(a) (b)

Figure 5: Microstructure and discretisation of the RVEs. (a) RVE-25i; and (b) RVE-5i.

Table 1: Quantity and dimensions of aggregates for each RVE.

<table>
<tr><th>RVE</th><th>Total number of aggregates</th><th colspan="3">Aggregate diameter/aggregate quantity</th><th>Aggregates volume fraction</th></tr>
<tr><td rowspan="2">RVE-5i</td><td rowspan="2">5</td><td>33 m</td><td colspan="2">27 mm</td><td rowspan="2">40.00%</td></tr>
<tr><td>4</td><td colspan="2">1</td></tr>
<tr><td rowspan="2">RVE-25i</td><td rowspan="2">25</td><td>19 mm</td><td>15 mm</td><td>12 mm</td><td rowspan="2">40.61%</td></tr>
<tr><td>5</td><td>6</td><td>14</td></tr>
</table>

Table 2: Discretisation of RVES.

| RVE | Discretisation | | | | | |
|---|---|---|---|---|---|---|
| | No. of nodes | No. of cells | No. of boundary elements | No. of inclusion interface nodes | No. of void interface nodes | No. of cohesive-contact elements |
| RVE-5i | 403 | 520 | 68 | 192 | 32 | 96 |
| RVE-25i | 721 | 1116 | 129 | 496 | 96 | 80 |

The aggregates are considered elastics with Young's modulus ($E$) = 40 GPa and Poisson's ratio ($v$) = 0.3, while the mortar matrix is governed by Mohr–Coulomb criterion, with friction angle equal to 4°, $E$ = 23 GPa, $v$ = 0.2. The plastic curve of the Mohr–Coulomb $((\bar{\varepsilon}_{p,}\sigma_y))$ where $\bar{\varepsilon}_p$ is the effective plastic strain and $\sigma_y$ the yield stress, is adopted, respectively, for RVE-25i and RVE-5i (0; 11.0) and (0.22; 70); (0; 11.7) and (0.22; 46). For RVE-25i the cohesive contact finite elements are defined only around the bigger aggregates, whereas for RVE-5i they are defined around all aggregates. Their parameters are defined as: $\beta_c$ = 0.707; $\sigma_c$ = 2 MPa; $\delta c$ = 0.0568 mm; $\lambda_p$ = 200,000. The value $\beta_c$ is related to the sliding and normal opening displacements (the higher its value, the greater sliding displacement importance); $\sigma_c$ is the maximum cohesive normal traction; $\delta c$ is a characteristic opening displacement; the penalty factor $\lambda_p$ is used in the contact model and it is related to the stiffness between triangular cells.

Fig. 6 shows the displacement $u_n$ at node 21 (see Fig. 4) throughout the load incremental process, comparing the results to the model where the finite element formulation discussed in Pituba et al. [9] models the RVE. 'Plate 1' is the plate where RVE-5i is used while RVE-25i is considered in 'Plate 2'. Fig. 7 compares the experimental curve to the numerical results.

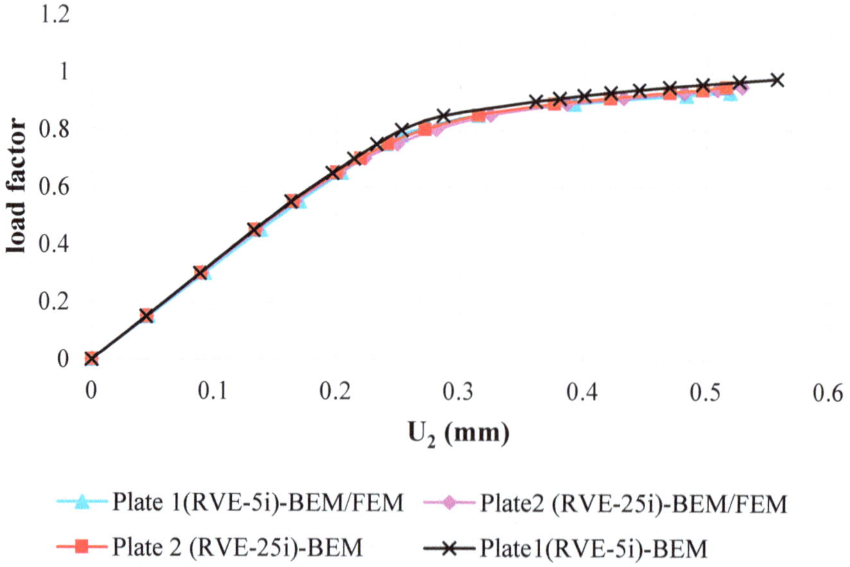


Figure 6: Displacement at the plate point 21 throughout the load incremental.

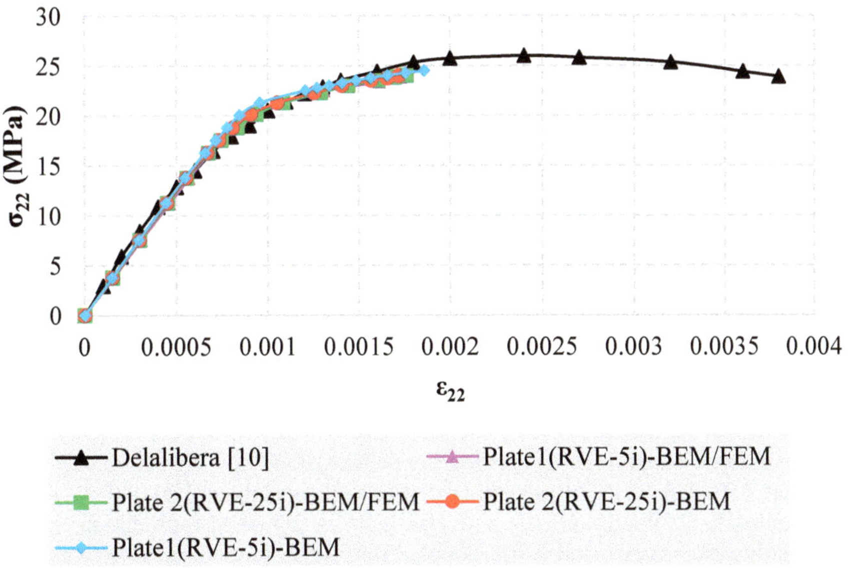


Figure 7: Stress versus strain in $x_2$ direction, comparing the numerical results to the experimental ones given in Delalibera [10].

We can observe that the plates present similar mechanical behaviour what means that a simple RVE can reproduce good results when only compression loads are considered. But it is important to stress that for more complex types of loading the RVEs will not present similar results, as the fracture process plays an important role when tension or shear stresses are considered. Besides, we observe that the BEM/FEM model presents similar results to the model where only BEM is considered. But the BEM model has reduced computational effort (see Table 3, where $t_{BEM}$ refers to the computational time of BEM model and $t_{BEM/FEM}$ to the computational time of BEM/FEM model). Table 3 also shows, as expected, that the computational time $t_{(plate\ 2)}$ related to 'Plate 2' is much bigger than the time related to 'Plate 1' $t_{(plate\ 1)}$.

Table 3: Comparison of the computational effort.

| **Plate 1 (RVE-5i)** $t_{BEM}/t_{BEM/FEM}$ | **Plate 2 (RVE-25i)** $t_{BEM}/t_{BEM/FEM}$ | **BEM** $t_{plate\ 1}/t_{plate\ 2}$ | **BEM/FEM** $t_{plate\ 1}/t_{plate\ 2}$ |
|---|---|---|---|
| 0.566 | 0.5665 | 0.1787 | 0.1789 |

Fig. 7 shows that the numerical results compare well to the experimental test, showing that the model is robust and efficient.

ACKNOWLEDGEMENTS
The authors want to thank CNPq (National Council for Scientific and Technological Development) (grant numbers 302809/2020-1) and FAPEG (Goiás Research Foundation) (grant number 202410267000680) for the financial support.

REFERENCES

[1] Peric, D., de Souza Neto, E.A., Feijóo, R., Partovi, M. & Molina, A.C., On micro-to-macro transitions for multiscale analysis of heterogeneous materials: Unified variational basis and finite element implementation. *Int. J. Numer. Methods Engrg.*, **87**, pp. 149–170, 2011.

[2] Somer, D.D., de Souza Neto, E.A., Dettmer, W.G. & Peric, D., A sub-stepping scheme for multi-scale analysis of solids. *Comput. Methods Appl. Mech. Engrg.*, **198**, pp. 1006–1016, 2009.

[3] Fernandes, G.R., Pituba, J.J.C. & de Souza Neto, E.A., FEM/BEM formulation for multi-scale analysis of stretched plates. *Eng. Anal. Bound. Elem.*, **54**, pp. 47–59, 2015.

[4] Fernandes, G.R., Crozariol, L.H.R., Furtado, A.S. & Santos, M.C., A 2D boundary element formulation to model the constitutive behaviour of heterogeneous microstructures considering dissipative phenomena. *Eng. Anal. Bound. Elem.*, **99**, pp. 1–22, 2019. https://doi.org/10.1016/j.enganabound.2018.10.018.

[5] Fernandes, G.R., Marques, M.J., Vieira, J.F. & Pituba, J.J.C., A RVE formulation by the boundary element method considering phase debonding. *Eng. Anal. Bound. Elem.*, **104**, pp. 259–276, 2019. https://doi.org/10.1016/j.enganabound.2019.03.018.

[6] Fernandes, G.R., Pontes, G.B.S., Oliveira, V.N., A 2D BEM formulation considering dissipative phenomena and a full coupled multiscale modelling. *Eng. Anal. Bound. Elem.*, **119**, pp. 25–43, 2020. https://doi.org/10.1016/j.enganabound.2020.07.004.

[7] Fernandes, G.R. & de Souza Neto, E.A., Self-consistent linearization of non-linear BEM formulations with quadratic convergence. *Computational Mechanics*, **52**, pp. 1125–1139, 2013.

[8] Silva, M.J.M., Pitaluga, C.G., Fernandes, G.R. & Pituba, J.J.C., Meso-scale modeling of the compressive mechanical behavior of concrete by a RVE-based BEM formulation. *Mechanics of Advanced Materials and Structures*, **1**, pp. 1–20, 2022.

[9] Pituba, J.J.C., Fernandes, G.R. & Souza Neto, E.A., Modeling of cohesive fracture and plasticity processes in composite microstructures. *Journal of Engineering Mechanics*, **142**, 04016069, 2016. https://doi.org/10.1061/(ASCE)EM.1943-7889.0001123.

[10] Delalibera, R.G., Theorical and experimental analysis of reinforced concrete beams with confinement reinforcement. Master's thesis, University of Sao Paulo, São Paulo, Brazil, 2002. (In Portuguese.)

# MODELLING OF CONCRETE USING THE BOUNDARY ELEMENT METHOD AND HOMOGENISATION TECHNIQUE

MARIA JULIA M. SILVA, CALEB G. PITALUGA, GABRIELA R. FERNANDES & JOSÉ J. C. PITUBA
Civil Engineering Department, Federal University of Catalão (UFCAT), Brazil

ABSTRACT
A 2D boundary element method (BEM) is used to simulate the mechanical behaviour of concrete within a multiscale framework considering the concept of representative volume element (RVE). The problem consists of solving the equilibrium problem in terms of displacement fluctuations at meso-scale of the material. For that, numerical analyses are performed using a consistent tangent operator for BEM formulation. A step-by-step analysis is detailed described for the parametric identification of the variables involved in a uniaxial compression test experimentally tested. The Mohr–Coulomb model is adopted to simulate the mortar matrix while elastic behaviour is considered for aggregates. Besides, the microcracking process at interfacial transition zone is modelled by defining additional cohesive-contact finite elements on interfaces between mortar and aggregates. The proposed numerical model shows its capability to perform this kind of analysis comparing with experimental test.
*Keywords: multiscale modelling, homogenisation, RVE, boundary elements, concrete.*

## 1 INTRODUCTION

The understanding of the mechanical properties of materials is crucial to obtain the best performance of them. If numerical analyses are associate with experimental ones, a reduced financial cost can be achieved to develop materials to particular uses. The present work focuses on concrete, the most used material in construction worldwide. In order to analyse the mechanical properties of concrete, the compressive strength is one of the most important parameters and deserves special attention. In fact, that parameter is widely used in specifying concrete under technical standards. Several studies have been developed considering concrete as a composite material formed by materials with different mechanical properties and arrangements leading to complex dissipative phenomena. Therefore, studies that carrying out the aspects of the mechanical behaviour of concrete are welcome.

In this context, the modelling of the mechanical behaviour of concrete using the representative volume element (RVE) concept and homogenisation technique has shown its efficiency and accuracy [1]–[3]. This approach provides macro constitutive models using simple constitutive models on the microstructural level. Besides, to solve the equilibrium problem of this kind of formulation, several numerical methods have been used, mainly finite element method (FEM) [4]–[8]. However, few works deal with this problem using boundary element methods (BEM) [6], [9]–[11]. In the present work, a BEM formulation is used to simulate the mechanical behaviour of concrete considering interfacial transition zone (ITZ) modelled by cohesive fracture elements and adopting a standard Mohr–Coulomb model and elastic behaviour for mortar matrix and aggregates, respectively. Besides, porous are inserted in the mortar matrix. In other works (Fernandes et al. [10], [11]), BEM has proved its efficiency when compared to others numerical models leading to reliable results but with smaller computational effort.

Finally, our goal is to show the potentialities of BEM when applied in practical situations as presented as in this work dealing with parametric identification of a concrete experimentally tested.

WIT Transactions on Engineering Sciences, Vol 136, © 2024 WIT Press
www.witpress.com, ISSN 1743-3533 (on-line)
doi:10.2495/BE470131

## 2 MULTISCALE BOUNDARY ELEMENT APPROACH

A BEM formulation detailed discussed in Fernandes et al. [10] and Fernandes and Souza Neto [12] has been used on the development of a multiscale approach based on the RVE concept. In this multiscale boundary element approach (MBEA), the 2D-dimensional plate problem is defined to model the RVE on the microstructure level. Also, the BEM equations are used to perform the analysis related to macrocontinuum. The communication between scales is made by computational homogenisation techniques detailed in Santos et al. [7], Santos and Pituba [8] and Borges and Pituba [13].

In case of small strain problems considering plasticity phenomenon, the total strain $\dot{\varepsilon}_{ij}$ is split into its elastic and plastic parts,$\dot{\varepsilon}^{e}_{ij}$ and $\dot{\varepsilon}^{p}_{ij}$, respectively. On the other hand, when phase debonding fracture processes are considered, the total strain $\dot{\varepsilon}_{ij}$ is split into the continuum $\dot{\bar{\varepsilon}}_{ij}$ and the fracture strains $\dot{\varepsilon}^{cf}_{ij}$. Therefore, the strain field in the RVE is defined as [14]:

$$\dot{\bar{\varepsilon}}_{ij} + \dot{\varepsilon}^{cf}_{ij} = \dot{\varepsilon}^{e}_{ij} + \dot{\varepsilon}^{p}_{ij}. \tag{1}$$

The Betti's theorem is used to define the integral representation of displacement. For a particular region of the plate composed of different materials, the Betti´s theorem can be written as follows (see more details in Fernandes et al. [10], [11] and Silva et al. [14]):

$$\int_{\Omega} \varepsilon^{*}_{kij}\left(\dot{N}_{ij} + \dot{N}^{0}_{ij}\right) d\Omega = \sum_{m=1}^{N_S} \left[ \frac{\overline{\overline{E}}_m}{\overline{\overline{E}}} \frac{\nu_m}{\nu} \int_{\Omega_m} \dot{\bar{\varepsilon}}_{ij} N^{*}_{kij} d\Omega + \overline{E}_m \left(1 - \frac{\nu_m}{\nu}\right) \int_{\Omega_m} \dot{\bar{\varepsilon}}_{ij} \varepsilon^{*}_{kij} d\Omega_m \right]$$

$$k,\, i,\, j = 1, 2, \tag{2}$$

in which terms with the superscript * relate to the fundamental problem, $\overline{\overline{E}}_m = \frac{\bar{E}_m}{(1-\nu_{m^2})}$, $\bar{E}_m = E_m t$, $E$ is the Young's modulus; $\nu$, the Poisson's ratio; $t$, the plate thickness; $\delta_{ij}$, the Kronecker's delta; $N_s$ is the number of subregions; $\dot{N}_{ij}$ is the membrane force obtained from the stress tensor rate $\dot{\sigma}_{ij}$, obtained by the plastic criterion; inelastic forces $\dot{N}^{0}_{ij}$ are related to plasticity and phase debonding phenomena and k, the fundamental load direction; $\nu$, $\overline{E}$, and all fundamental values refer to the subregion in which the collocation point is placed.

On the other hand, the Hill–Mandel Principle is used to define the RVE equilibrium problem. This Principle guarantees the energy consistency between scales. After several mathematical manipulations and the RVE discretisation into $N_{cel}$ cells, the principle leads to the equilibrium eqn (3) considering the displacement field obtained from the macroscopic strain $\varepsilon(x)$ imposed to the RVE boundary and the displacement fluctuation field $\tilde{u}_{\mu}$.

$$R_F = \sum_{e=1}^{N_{cel}} B_e^T \Delta\sigma_{\mu}^{e(i)} t\, A_e = \sum_{e=1}^{N_{cel}} [B]_e^T t \left[D_{\mu}^{e}\right] (\{\Delta\varepsilon\} + [B]\{\Delta\tilde{u}\})_e A_e = 0, \tag{3}$$

in which $\left[D_{\mu}^{e}\right]$ is the constitutive tensor related to the cell $e$, $B_e$, the cell strain-displacement matrix, $A_e$, the cell area; and $N_{\mu}^{e} = t\sigma_{\mu}^{e}$, the cell membrane force increment vector, considered constant over the cell.

## 3 NUMERICAL APPLICATION: PARAMETRIC IDENTIFICATION OF CONCRETE USING MBEA

In this section is presented a parametric identification of a concrete experimentally tested by Delalibera [15] using the MBEA described in Section 2. Accordingly with experiment tests carried out by Delalibera [15], 40.61% of aggregate volume fraction has been adopted for the RVE distributed as five aggregates with 19 mm of diameter, six aggregates with 15 mm and 14 aggregates with 12 mm. The RVE adopted is shown in Fig. 1. Also, it is adopted an elastic behaviour for the aggregates, Mohr–Coulomb criterion to model the mortar matrix, both discretised by triangular element cells while a contact and cohesive fracture model is used in the ITZ modelled by interface elements (see Table 1). In all analyses, a macrostrain tensor is imposed to the RVE ($\varepsilon_x = -0.003$; $\varepsilon_y = 0.0006$ and $\varepsilon_{xy} = 0$), which is divided into 20 increments adopting periodic fluctuations to the RVE boundary as discussed in Santos et al. [7], Santos and Pituba [8], Fernandes et al. [10] and Borges and Pituba [13]. The goal of this section is to evidence the potentialities of the MBEA in representing the concrete mechanical behaviour submitted to uniaxial compression stress state capturing the major features of this material when compared to the experimental curve presented in Delalibera [15].

(a) (b) (c)

Figure 1: RVE discretisation. (a) RVEa with 634 cells; (b) RVEb with 564 cells; and (c) RVEc with 1276 cells.

Table 1: Properties of the mortar matrix, aggregates and ITZ.

| Phase | Properties | | | | |
|---|---|---|---|---|---|
| | **ν** | **E (GPa)** | **Adopted constitutive model** | **Hardening points** ($\bar{\varepsilon}_p, \sigma_y(MPa)$) | **Friction angle (ϕ)** |
| Mortar matrix | 0.2 | 23 | Mohr–Coulomb | (0;11) (0.22;40) | 4° |
| Aggregates | 0.3 | 40 | Elastic | – | – |
| | **$\beta_c$** | **$\sigma_c$ (MPa)** | **$\delta_c$ (mm)** | **$\lambda_p$** | |
| ITZ | 0.707 | 1.00 | 0.0568 | 200,000.00 | |

First of all, three different discretisation have been used to model the problem in order to achieve the objectivity of the numerical response. Table 2 contains information about the discretisation. Fig. 2 shows the homogenised stress in the *x* direction ($\sigma_x^{hom}$) versus imposed macrostrain in the *x* direction curves obtained by different meshes. As it can see, the meshes show the same results. In particular, the poorest mesh is then considered for further analyses. Note that the using of cohesive and contact finite elements have been adopted only on the interfaces between the matrix and bigger aggregates overcoming instability problems inherent to this kind of element.

Table 2: Information on the discretisation of the RVEs.

| RVE | Number of nodes, elements and cells | | | | | |
|---|---|---|---|---|---|---|
| | Nodes | Cells | Boundary elements | Interface nodes | Interface elements | Cohesive and contact elements |
| RVEa | 412 | 634 | 68 | 280 | 220 | 60 |
| RVEb | 564 | 882 | 84 | 344 | 264 | 80 |
| RVEc | 1276 | 2214 | 136 | 520 | 420 | 100 |

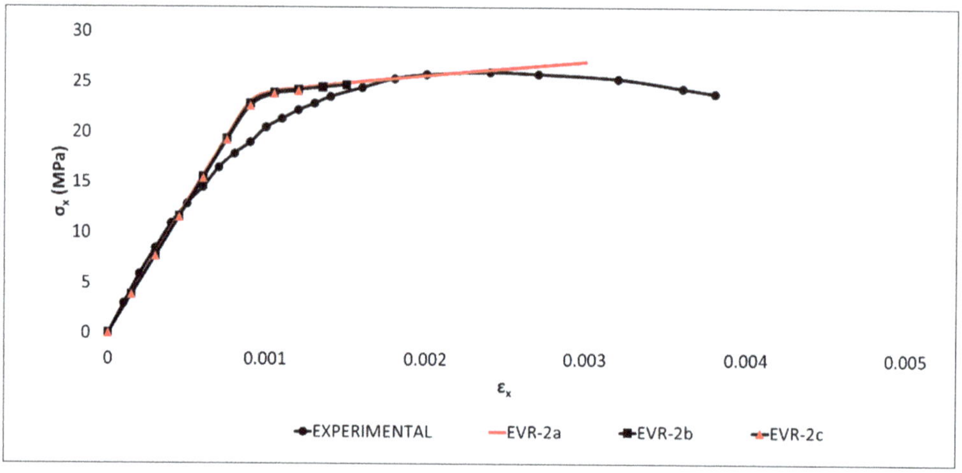


Figure 2: Homogenised stress and imposed macrostrain strain curves for RVEa, RVEb and RVEc.

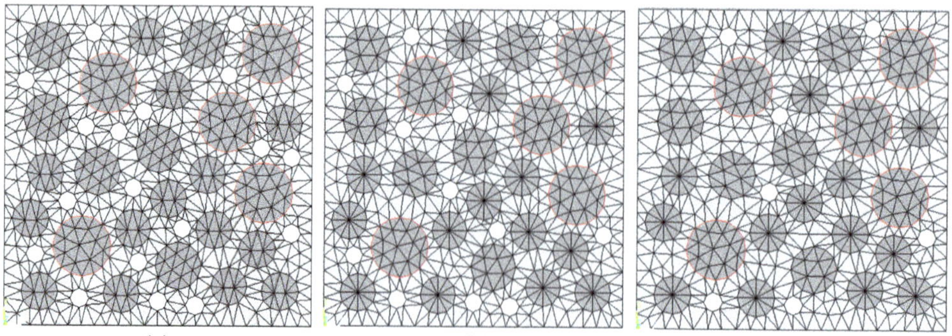

(a) (b) (c)

Figure 3: RVEs with different porosity fractions. (a) 5%; (b) 3.34%; and (c) 2%. Red lines around the bigger aggregates indicate the use of contact and cohesive fracture elements.

To improve the numerical response around 15 MPa, when the microcracking process starts to play an important role on the dissipative phenomena, the porosity of the mortar matrix must be considered. Besides, this kind of porosity is observed experimentally. Therefore, a study is carried out considering the insertion of porous following the proportions presented in Fig. 3. Fig. 4 shows the results of the RVE-0% (former RVE described in Fig. 2), RVE-2%, RVE-3.34% and RVE-5%. Note that the results are improved in the pre-peak region, but a strength reduction is observed, as expected. As it can see, the RVE-3.34% represents better the experimental response.

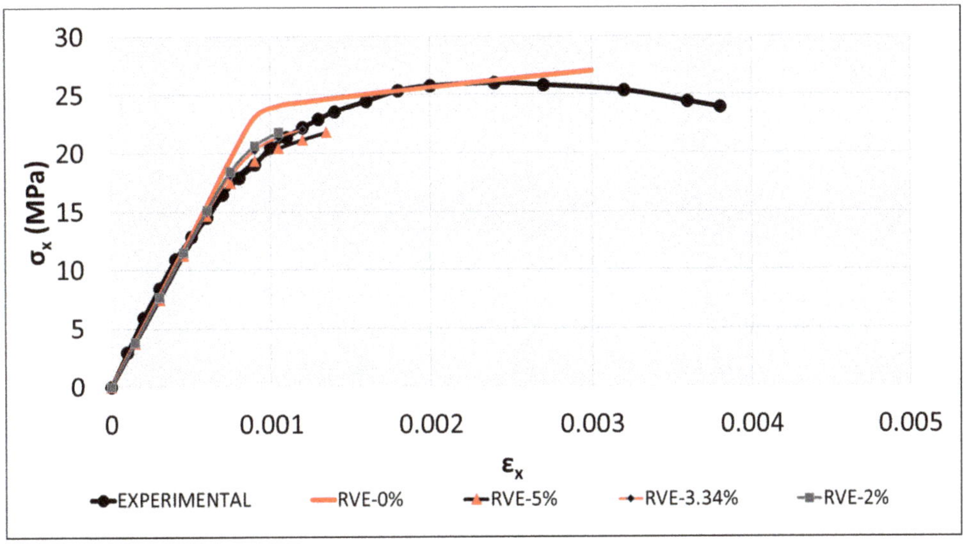


Figure 4: Homogenised stress in the x direction versus imposed macrostrain for RVE-0%, RVE-2%, RVE-3.34% and RVE-2%.

Finally, the RVE-3.34% model leads to a reduced strength of the homogenised material. In order to overcome this issue a change of isotropic hardening curve related to Mohr–Coulomb model for mortar matrix is made (see Table 1). The new parameters of the isotropic hardening curve ($\bar{\varepsilon}_p, \sigma_y(MPa)$) are given by (0;11) and (0.22;70). The new RVE is called RVE-3.34%F. Fig. 5 shows results for the RVE-3.34% and the new one RVE-3.34%F. Note that the homogenised macroscopic model predicts the experimental response quite well until the peak stress zone obtaining the experimentally observed concrete strength. Note that a different RVE modelling could be proposed in which microcracks could be created and propagated inside the mortar matrix to capture the softening behaviour evidenced in pos-peak region.

## 4 CONCLUSIONS

In the present work, a 2D-dimensaional formulation is presented using the BEM and computational homogenisation technique within a multiscale approach. In this work, the MBEA has been used in numerical analyses just on the mesoscale level considering RVE of concrete previously experimental tested by Delalibera [15]. The MBEA has proved to be a feasible alternative to perform analyses of concrete capturing the ultimate compressive

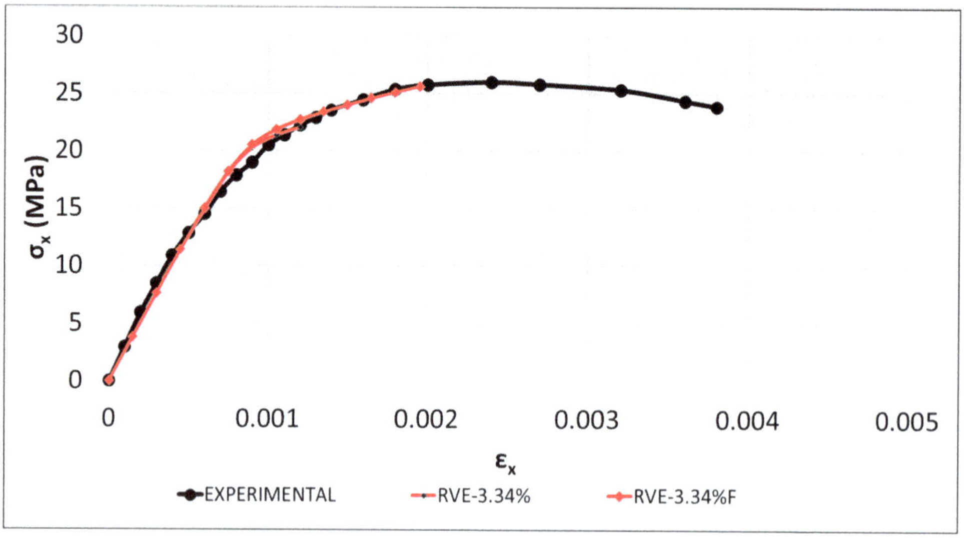


Figure 5: Homogenised stress in the x direction versus imposed macrostrain for RVE-3.34% and RVE-3.34%F.

strength of the material as well as the major features of its mechanical behaviour. Besides, the porosity in the mortar matrix must be considered in the homogenised macroscopic model despite the reduction of the strength evidenced by the model. In fact, this reduction is experimented by the material and the porosity has important role in this phenomenon.

## ACKNOWLEDGEMENTS

This work was financially supported by CNPq (National Council for Scientific and Technological Development) (grant numbers 305927/2023-0 and 303863/2023-4), the CAPES Foundation (Ministry of Education of Brazil) (grant number 88887.485859/2020-00) and FAPEG (Goias Research Foundation) (grant number 202410267000679).

## REFERENCES

[1] Karamnejad, A., Nguyen, V.P. & Sluys, L.J., A Multi-scale rate dependent crack model for quasi-brittle heterogeneous materials. *Engineering Fracture Mechanics*, **104**, pp. 96–113, 2013. http://doi.org/10.1016/j.engfracmech.2013.03.009.

[2] Pituba, J.J.C. & Souza Neto, E.A., Modeling of unilateral effect in brittle materials by a mesoscopic scale approach. *Computers and Concrete*, **15**, pp. 735–758, 2015. https://doi.org/10.12989/cac.2015.15.5.735.

[3] Toro, S., Sánchez, P.J., Blanco, P.J., Souza Neto, E.A., Huespe, A.E. & Feijóo, R.A., Multiscale formulation for material failure accounting for cohesive cracks at the macro and micro scales. *International Journal of Plasticity*, **76**, pp. 75–110, 2016. http://doi.org/10.1016/j.ijplas.2015.07.001.

[4] Peric, D., Souza Neto, E.A., Feijóo, R.A., Partovi, M. & Molina, A.J.C., On micro to macro transitions for multi scale analysis of non linear heterogeneous materials: Unified variational basis and finite element implementation. *Numerical Methods in Engineering*, **87**, pp. 149–170, 2011.

[5] Somer, D.D., Souza Neto, E.A., Dettmer, W.G. & Peric, D., A sub-stepping scheme for multi-scale analysis of solids. *Computer Methods in Applied Mechanics and Engineering*, **198**, pp. 1006–1016, 2009.

[6] Fernandes, G.R., Marques, M.J., Vieira, J.F. & Pituba, J.J.C., A RVE formulation by the boundary element method considering phase debonding. *Engineering Analysis with Boundary Elements*, **104**, pp. 259–276, 2019. https://doi.org/10.1080/17797179.2017.1379863.

[7] Santos, W.F., Fernandes, G.R. & Pituba, J.J.C., Analysis of the influence of plasticity and fracture processes on the mechanical behavior of metal matrix composites microstructures. *Revista Materia*, **21**, pp. 577–598, 2016. https://doi.org/10.1590/S1517-707620160003.0056.

[8] Santos, W.F. & Pituba, J.J.C., Yield surfaces of material composed of porous and heterogenous microstructures considering phase debonding. *Latin American Journal of Solids and Structures*, **14**, pp. 1387–1415, 2017. https://doi.org/10.1590/1679-78253776.

[9] Rodríguez, R.Q., Moura, L.S., Galvis, A.F., Albuquerque, E.L., Tan, C.L. & Sollero, P., Multi-scale dynamic failure analysis of 3D laminated composites using BEM and MCZM. *Engineering Analysis with Boundary Elements*, **104**, pp. 94–106, 2019. https://doi.org/10.1016/j.enganabound.2019.03.017.

[10] Fernandes, G.R., Furtado, A.S., Pituba, J.J.C. & de Souza Neto, E.A., Multiscale Analysis of structures composed of metal matrix composites considering phase debonding. *Journal of Multiscale Modelling*, **8**, 1740004, 2017.

[11] Fernandes, G.R., Crozariol, L.H.R., Furtado, A.S. & Santos, M.C., A 2D boundary element formulation to model the constitutive behaviour of heterogeneous microstructures considering dissipative phenomena. *Engineering Analysis with Boundary Elements*, **99**, pp. 1–22, 2019. https://doi.org/10.1016/j.enganabound.2018.10.018.

[12] Fernandes, G.R. & Souza Neto, E.A., Self-consistent linearization of non-linear BEM formulations with quadratic convergence. *Computational Mechanics*, **52**, pp. 1125–1139, 2013.

[13] Borges, D.C. & Pituba, J.J.C., Homogenized damage model for brittle materials. *Mechanics of Advanced Materials and Structures*, 2023. https://doi.org/10.1080/15376494.2023.2258365.

[14] Silva, M.J.M., Pitaluga, C.G., Fernandes, G.R. & Pituba, J.J.C., Meso-scale modelling of the compressive mechanical behavior of concrete by a RVE-based BEM formulation. *Mechanics of Advanced Materials and Structures*, 2022. https://doi.org/10.1080/15376494.2022.2144974.

[15] Delalibera, R.G., Theorical and experimental analysis of reinforced concrete beams with confinement reinforcement. Master's thesis, University of Sao Paulo, São Paulo, Brazil, 2002.

# Author index

Milton Keynes UK
Ingram Content Group UK Ltd.
UKHW051926031224
451896UK00002B/5

9 781784 664930